ANALYTIC GEOMETRY

BY

LEWIS PARKER SICELOFF

GEORGE WENTWORTH

AND

DAVID EUGENE SMITH

GINN AND COMPANY

BOSTON · NEW YORK · CHICAGO · LONDON
ATLANTA · DALLAS · COLUMBUS · SAN FRANCISCO

The Athenæum Press
GINN AND COMPANY · PRO-
PRIETORS · BOSTON · U.S.A.

PREFACE

This book is written for the purpose of furnishing college classes with a thoroughly usable textbook in analytic geometry. It is not so elaborate in its details as to be unfitted for practical classroom use; neither has it been prepared for the purpose of exploiting any special theory of presentation; it aims solely to set forth the leading facts of the subject clearly, succinctly, and in the same practical manner that characterizes the other textbooks of the series.

It is recognized that the colleges of this country generally follow one of two plans with respect to analytic geometry. Either they offer a course extending through one semester or they expect students who take the subject to continue its study through a whole year. For this reason the authors have so arranged the work as to allow either of these plans to be adopted. In particular it will be noted that in each of the chapters on the conic sections questions relating to tangents to the conic are treated in the latter part of the chapter. This arrangement allows of those subjects being omitted for the shorter course if desired. Sections which may be omitted without breaking the sequence of the work, and the omission of which will allow the student to acquire a good working knowledge of the subject in a single half year are as follows: 46–53, 56–62, 121–134, 145–163, 178–197, 225–245, and part or all of the chapters on solid geometry. On the other hand, students who wish that thorough foundation in analytic geometry which should precede the study of the higher branches of mathematics are urged to complete the entire book, whether required to do so by the course of study or not.

This book is intended as a textbook for a course of a full year, and it is believed that many of the students who study the subject for only a half year will desire to read the full text. An abridged edition has been prepared, however, for students who study the subject for only one semester and who do not care to purchase the larger text.

It will be observed that the work includes two chapters on solid analytic geometry. These will be found quite sufficient for the ordinary reading of higher mathematics, although they do not pretend to cover the ground necessary for a thorough understanding of the geometry of three dimensions.

It will also be noticed that the chapter on higher plane curves includes the more important curves of this nature, considered from the point of view of interest and applications. A complete list is not only unnecessary but undesirable, and the selection given in Chapter XII will be found ample for our purposes.

CONTENTS

GREEK ALPHABET

The use of letters to represent numbers and geometric magnitudes is so extensive in mathematics that it is convenient to use the Greek alphabet for certain purposes. The Greek letters with their names are as follows:

A	α	alpha	N	ν	nu
B	β	beta	Ξ	ξ	xi
Γ	γ	gamma	O	o	omicron
Δ	δ	delta	Π	π	pi
E	ϵ	epsilon	P	ρ	rho
Z	ζ	zeta	Σ	σ, ς	sigma
H	η	eta	T	τ	tau
Θ	θ	theta	Υ	υ	upsilon
I	ι	iota	Φ	ϕ	phi
K	κ	kappa	X	χ	chi
Λ	λ	lambda	Ψ	ψ	psi
M	μ	mu	Ω	ω	omega

ANALYTIC GEOMETRY

CHAPTER I

INTRODUCTION

1. Nature of Algebra. In algebra we study certain laws and processes which relate to number symbols. The processes are so definite, direct, and general as to render a knowledge of algebra essential to the student's further progress in the study of mathematics.

As the student proceeds he may find that he has forgotten certain essential facts of algebra. Some of the topics in which this deficiency is most frequently felt are provided in the Supplement, page 283.

2. Nature of Elementary Geometry. In elementary geometry we study the position, form, and magnitude of certain figures. The general method consists of proving a theorem or solving a problem by the aid of certain geometric propositions previously considered. We shall see that analytic geometry, by employing algebra, develops a much simpler and more powerful method.

It is true that elementary geometry makes a little use of algebraic symbols, but this does not affect the general method employed.

3. Nature of Trigonometry. In trigonometry we study certain functions of an angle, such as the sine and cosine, and apply the results to mensuration.

The formulas of trigonometry needed by the student of analytic geometry will be found in the Supplement.

4. Nature of Analytic Geometry. The chief features of analytic geometry which distinguish it from elementary geometry are its method and its results. The results will be found as we proceed, but the method of procedure may be indicated briefly at once. This method consists of indicating by algebraic symbols the position of a point, either fixed or in motion, and then applying to these symbols the processes of algebra. Without as yet knowing how this is done, we can at once see that with the aid of all the algebraic processes with which we are familiar we shall have a very powerful method for exploring new domains in geometry, and for making new applications of mathematics to the study of natural phenomena.

The idea of considering not merely fixed points, as in elementary geometry, but also points in motion is borrowed from a study of nature. For example, a ball B thrown into the air follows a certain curve, and the path of a planet E about its sun is also a curve, although not a circle.

5. Point on a Map. The method by which we indicate the position of a point in a plane is substantially identical with the familiar method employed in map drawing. To state the position of a place on the surface of the earth we give in degrees the distance of the place east or west of the prime meridian, that is, the longitude of the place; and then we give in degrees the distance of the place north or south of the equator, that is, the latitude of the place.

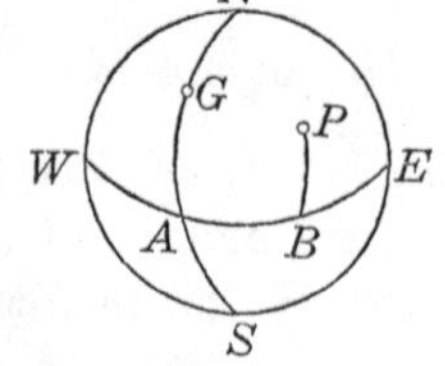

For example, if the curve $NGAS$ represents the prime meridian, a meridian arbitrarily chosen and passing through Greenwich, and if $WABE$ represents the equator, the position of a place P is determined if AB and BP are known. If $AB = 70°$ and $BP = 45°$, we say that P is 70° east and 45° north, or 70° E. and 45° N.

6. Locating a Point. The method used in map drawing is slightly modified for the purposes of analytic geometry. Instead of taking the prime meridian and the equator as lines of reference, we take any two intersecting straight lines to suit our convenience, these being designated as YY' and XX', and their point of intersection as O.

Then the position of any point P in the plane is given by the two segments OA and AP, AP being parallel to YY'.

If we choose a convenient unit of measurement for these two segments, the letters a and b may be taken to represent the numerical measures of the line segments OA and AP, or the distances from O to A and A to P respectively.

It is often convenient to draw PB parallel to AO and to let $BP = a$, as here shown. That is, we may locate the point P by knowing OA and AP, or OA and OB, or BP and AP, or BP and OB.

7. Convention of Signs. As in elementary algebra we shall consider the segment OA, or the distance a, in the figure next above, as positive when P is to the right of YY', and as negative when P is to the left of YY'. Similarly, we shall consider the segment AP, or the distance b, as positive when P is above XX', and as negative when P is below XX'.

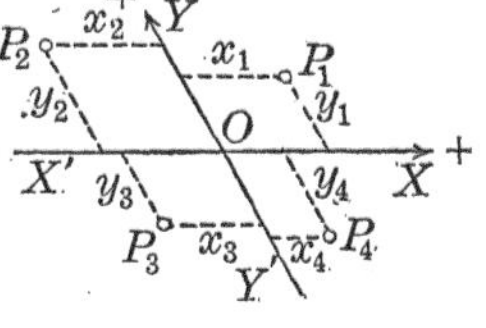

In this figure P_1 is determined by x_1 and y_1, P_2 by x_2 and y_2, P_3 by x_3 and y_3, and P_4 by x_4 and y_4. Furthermore, x_1, x_4, y_1, y_2 are positive, and x_2, x_3, y_3, y_4 are negative.

We shall hereafter speak of a line segment and its numerical measure as synonymous, and shall use the word *line* to mean a straight line, unless confusion is likely to arise, in which case the language will conform to the conditions which develop.

Exercise 1. Locating Points

Using 1 in. to represent 20°, and considering the surface of the earth as a plane, draw maps locating the following places:

1. 20° W., 40° N. **3.** 20° W., 20° S. **5.** 0° W., 0° N.

2. 20° E., 40° N. **4.** 20° E., 30° S. **6.** 10° W., 0° N.

Using $\frac{1}{8}$ in. as the unit of measure, draw any two intersecting lines and locate the following points:

7. $a=3, b=-2$. **9.** $a=-4, b=6$. **11.** $a=0, b=0$.

8. $a=4, b=5$. **10.** $a=-5, b=0$. **12.** $a=-7, b=-4$.

13. Given $x_1=-4$, $y_1=0$, locate the point P_1, and given $x_2=0$, $y_2=8$, locate the point P_2, and then draw the line P_1P_2. Find the pair of numbers which locate the mid point of P_1P_2.

14. In this figure OBA is an equilateral triangle whose side is 6 units in length. Taking XX' and YY' as shown, find the five pairs of numbers which locate the three vertices A, B, O and the mid points M_1 and M_2 of OA and AB respectively.

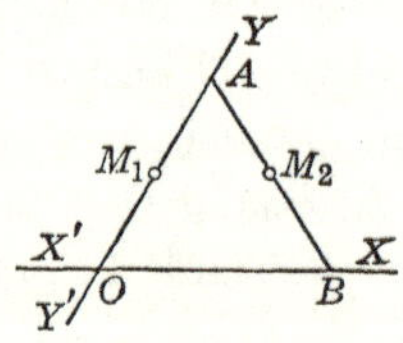

15. Given the equilateral triangle ABC in which $AB=6$. Taking XX' along the base, and YY' along the perpendicular bisector of the base, find the five pairs of numbers which locate the three vertices A, B, C and the mid points M_1 and M_2 of AB and BC respectively, as shown in the figure.

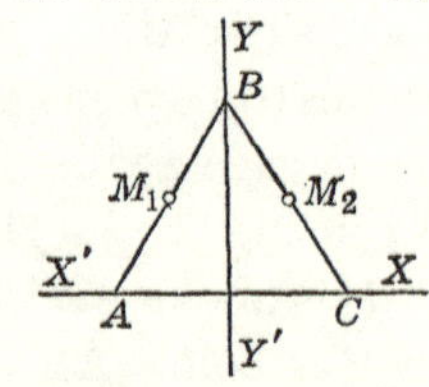

16. In this figure AB and CD are parallel, and BD is perpendicular to each. The length of BD is 7 units. Draw the circle as shown, tangent to the three lines. Taking XX' along CD, and YY' along BD, find the pairs of numbers which locate the center of the circle and the three points of tangency.

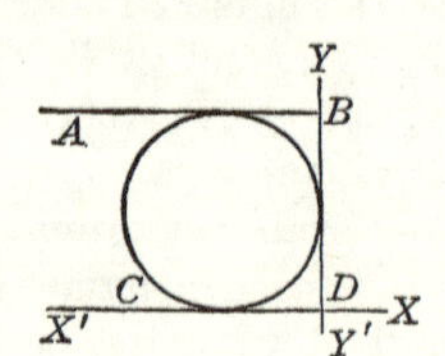

8. Definitions and Notation. In this figure the lines XX' and YY' are called respectively the *x axis* and the *y axis*, O is called the *origin*, OA, or a, is called the *abscissa* of P, and AP, or b, is called the *ordinate* of P. The abscissa and the ordinate of the point P taken together are called the *coordinates* of P.

Every pair of real numbers, a and b, are the coordinates of some point P in the plane; and, conversely, every point P in the plane has a pair of real coordinates.

When we speak of the point a, b, also written (a, b), we mean the point with a for abscissa and b for ordinate.

For example, the point $4, -3$, $(4, -3)$, or $P(4, -3)$ has the abscissa 4 and the ordinate -3, and the point $(0, 0)$ is the origin.

The method of locating a point described in § 6 is called the *method of rectilinear coordinates*.

Rectilinear coordinates are also called *Cartesian coordinates*, from the name of Descartes (Latin, Cartesius) who, in 1637, was the first to publish a book upon the subject.

When the angle XOY is a right angle, the coordinates are called *rectangular coordinates*; and when the angle XOY is oblique, the coordinates are called *oblique coordinates*.

Following the general custom, we shall employ only rectangular coordinates except when oblique coordinates simplify the work.

In the case of rectangular coordinates the axes divide the plane into four *quadrants*, XOY being called the first quadrant, YOX' the second, $X'OY'$ the third, and $Y'OX$ the fourth, as in trigonometry.

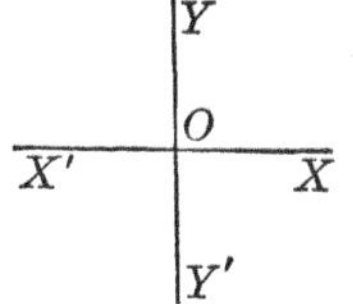

9. Coordinate Paper. For convenience in locating points, paper is prepared, ruled in squares of convenient size. This is called *coordinate paper*.

Exercise 2. Locating Points

1. Draw a pair of rectangular axes, use $\frac{1}{8}$ in. as the unit, and locate the following points: (4, 2), (2, 4), (– 2, 4), (– 4, 2), (– 4, – 2), (– 2, – 4), (2, – 4), (4, – 2).

2. Draw a pair of rectangular axes, use any convenient unit, and locate the following points: (–1, 4), (0, 2), (1, 1), (2, $\frac{1}{2}$), (3, $\frac{1}{4}$), (4, $\frac{1}{8}$), (5, $\frac{1}{16}$).

It should be noticed that in Exs. 1 and 2 the points lie on curves of more or less regularity.

3. Draw a pair of rectangular axes, use $\frac{1}{4}$ in. as the unit, and locate the following points: (5, 5), (5, – 5), (– 5, 5), (– 5, – 5), (0, 5), (5, 0), (– 5, 0), (0, – 5).

4. Draw a pair of rectangular axes, use any convenient unit, and locate the following points: (30, –10), (25, 17), (– 20, – 24), (–12, 20), (– 30, –10), (21, 28), (– 21, 28).

In each of the above four examples the drawing of the axes has been mentioned, and the unit also has been mentioned. Hereafter it will be understood that the axes are to be drawn and a convenient unit is to be taken. It is suggested, however, that the student hereafter use coordinate paper as described in § 9.

Locate each of the following pairs of points and calculate the distance between the points:

5. (0, 2), (0, 5). **8.** (0, 2), (0, – 5). **11.** (4, 0), (8, 0).

6. (4, 0), (4, 6). **9.** (0, – 2), (0, 5). **12.** (– 6, 2), (4, 2).

7. (6, 2), (6, 8). **10.** (4, – 3), (4, 3). **13.** (–1, 0), (0, –1).

14. Locate the following points and calculate the distance from the origin to each point: (3, 4), (– 3, 4), (– 4, – 3), (2, – 5), (– 6, – 3), (5, 5).

In place of calculating square roots, the table of square roots given on page 284 in the Supplement may be used.

15. Locate the following points and calculate the distance of each from the origin: (8, 4), (– 8, 4), (– 8, – 4), (7, – 2).

16. Draw perpendiculars from $P_1(-8, -8)$ and $P_2(-2, -4)$ to both axes, and find the coordinates of the mid point of P_1P_2.

17. Draw perpendiculars from $P_1(5, 6)$ and $P_2(11, 8)$ to both axes, and find the coordinates of the mid point of P_1P_2.

18. In this figure it is given that $OA = 7$ and $OB = 4$. Find the coordinates of each vertex of the rectangle and of the intersection of the diagonals.

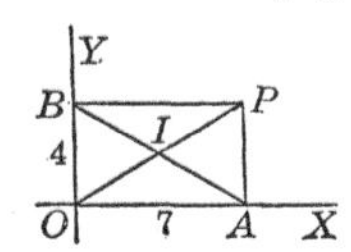

19. If the mid point of the segment P_1P_2 is (0, 0), and the coordinates of P_1 are a, b, find the coordinates of P_2.

20. In this figure it is given that BAO is a right triangle, M_1 is the mid point of OB, and M_2 is the mid point of AB. If the coordinates of B are 13, 11, find the coordinates of M_1 and M_2. From the result show that M_1M_2 is parallel to OA.

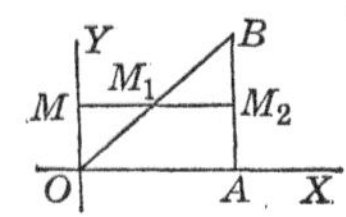

21. In the figure of Ex. 20 produce M_2M_1 to meet OY at M, and find the lengths of MM_2, M_1M_2, and OM_1.

22. Construct a circle through the three points $A(6, 0)$, $B(0, 8)$, and $O(0, 0)$, and find the coordinates of the center and the length of the radius.

23. Draw the bisector of the angle XOY and produce it through O. On this line locate a point whose abscissa is -4 and find the ordinate of the point.

24. A point $P(x, y)$ moves so that its abscissa is always equal to its ordinate, that is, so that x is always equal to y. Find the path of P and prove that your conclusion is correct.

25. Draw the path of a point which moves so that its ordinate always exceeds its abscissa by 1.

26. Describe the position of all points $P(x, y)$ which have the same abscissa.

For example, consider the points whose abscissas are all equal to 5.

27. Describe the position of all points $P(x, y)$ which have the same ordinate.

10. Lines and Equations. Having learned how to locate points by means of coordinates, we now turn to the treatment of lines, straight or curved.

For our present purposes, a line may conveniently be regarded as made up of all the points which lie upon it.

Consider, for example, the bisector AB of the angle XOY. Here we see that the coordinates of every point $P(x, y)$ on the line AB are equal; that is, in every case $x = y$. And conversely, if $x = y$, the point (x, y) is on AB.

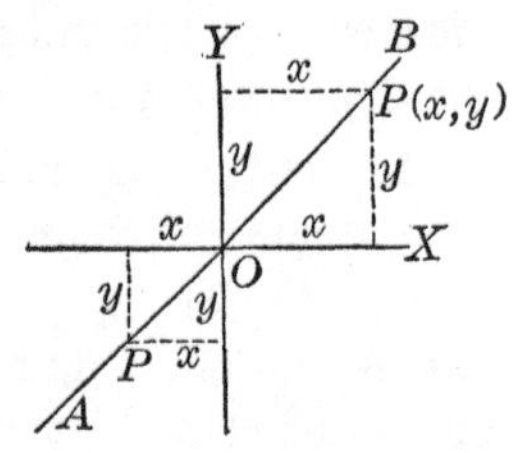

Hence the equation $x = y$ is true for all points on AB, and for no other points. We therefore say that the equation $x = y$ is the *equation of the line* AB, and refer to AB as the *graph*, or *locus*, *of the equation* $x = y$.

Again, consider the circle with center at the origin and with radius 5. If the point $P(x, y)$ moves along the circle, x and y change; but since x and y are the sides of a right triangle with hypotenuse 5, it follows that $x^2 + y^2 = 25$. This equation is true for all points on the circle and for no others; it is the equation of this particular circle, and the circle is the graph of the equation $x^2 + y^2 = 25$.

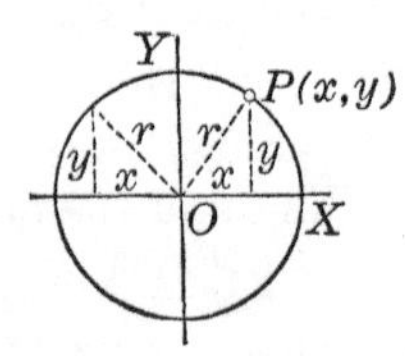

We therefore see that a certain straight line and a certain circle, which are geometric, are represented algebraically in our system of coordinates by the equations $x = y$ and $x^2 + y^2 = 25$ respectively.

Whenever an equation is satisfied by the coordinates of all points of a certain line, and by the coordinates of no other points, the line is called the *graph of the equation*, and the equation is called the *equation of the graph*.

11. Two Fundamental Problems. The notion of correspondence between graphs and equations gives rise to two problems of fundamental importance in analytic geometry:

1. *Given the equation of a graph, to draw the graph.*
2. *Given a graph, to find its equation.*

These problems, with the developments and applications to which they lead, form the subject matter of analytic geometry. A few examples illustrating the first of these problems will now be given, the second problem being reserved for consideration in Chapter III.

The word *locus* is often employed in place of the word *graph* above. So also is the word *curve,* which includes the straight line as a special case.

12. Nature of the Graph of an Equation. In general an equation in x and y is satisfied by infinitely many pairs of real values of x and y. Each of these pairs of values locates a point on the graph of the equation. The set of all points located by these pairs usually forms, as in the examples on page 8, a curve which contains few or no breaks, or, as we say, a curve which is *continuous* through most or all of its extent.

In dealing with graphs of equations in the remainder of this chapter, and again in Chapter III, we shall assume that the graphs are continuous except when the contrary is shown to be the case.

The study of the conditions under which the graph of an equation is not continuous, together with related topics, is a matter of considerable importance in the branch of mathematics known as the calculus.

In special cases it may happen that the graph of an equation consists of only a limited number of real points. For example, the equation $x^2 + y^2 = 0$ is satisfied only by $x = 0$, $y = 0$, and its graph is a single point, the origin.

13. Plotting the Graph of a Given Equation. If we have given an equation in x and y, such as $3y - 5x + 8 = 0$, we can find any desired number of points of its graph, thus:

When $x =$	-2	-1	0	1	2	3	4
then $y =$	-6	$-4\frac{1}{3}$	$-2\frac{2}{3}$	-1	$\frac{2}{3}$	$2\frac{1}{3}$	4

Each pair of values thus found satisfies the equation and locates a point of the graph. By plotting these points and connecting them by a curve we are led to infer, although we have not yet proved it, that the graph is a straight line.

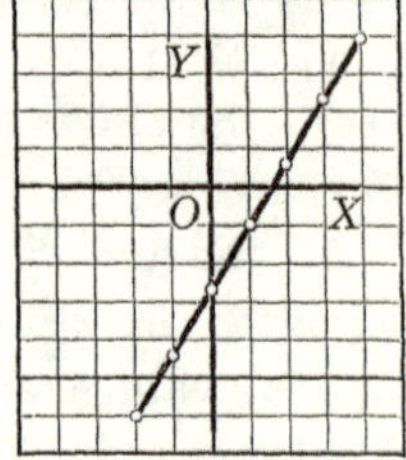

The above process is called *plotting the graph of the equation*, or simply *plotting the equation*.

As a second illustration of plotting an equation we may consider the case of $y = 1 + x - x^2$. Proceeding as before:

When $x =$	-3	-2	-1	0	1	2	3	4
then $y =$	-11	-5	-1	1	1	-1	-5	-11

Joining the points by a smooth curve, we have a curve known as the parabola, which is defined later.

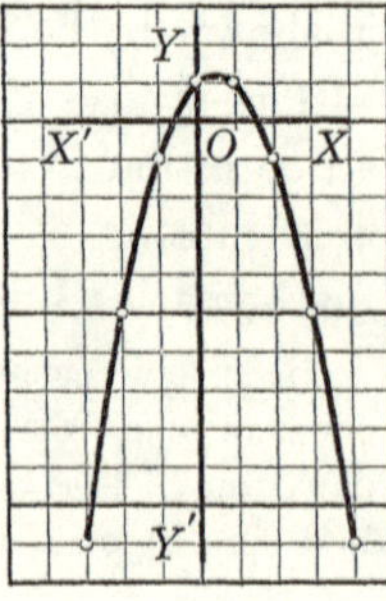

It may be mentioned at this time that a ball thrown in the air would follow a parabolic path if it were not for the resistance of the air; that the paths of certain comets are parabolas; that parabolic arches are occasionally used by engineers and architects; and that the cables which support suspension bridges are usually designed as parabolas.

Exercise 3. Graphs of Equations

1. Plot the equation $y - 2x + 2 = 0$.

2. Plot the equations $3x + 2y = 6$ and $3x + 2y = 12$.

3. Plot the equations $4x - 3y = 10$ and $3x + 4y = 12$.

4. Plot the equation $y = x^2$, taking $x = -4, -2, 0, 2, 4$.

5. Plot the equation $y = +\sqrt{x}$, locating the points determined by giving to x the values 0, 1, 4, 9, and 16. By examining the graph, determine approximately $\sqrt{6}$, $\sqrt{12}$, and $\sqrt{14}$.

The value of each square root should be estimated to the nearest tenth. The estimates may be checked by the table on page 284.

6. The equation $A = \pi r^2$ gives the area of a circle in terms of the radius. Using an A axis and an r axis as here shown, plot this equation, taking $\pi = \frac{22}{7}$, and assigning to r the values 0, 1, 2, 3. Estimate from the graph the change in area from $r = 1$ to $r = 2$, and from $r = 2$ to $r = 3$.

Why were no negative values of r suggested?

7. From each corner of a sheet of tin 8 in. square there is cut a square of side x inches. The sides are then bent up to form a box. Knowing that the volume V is the product of the base and height, express V in terms of x. Using a V axis and an x axis plot the equation, taking $x = 0, 1, 2, 3, 4$. From the graph, determine approximately the value of x which seems to give the greatest volume to the box. If your estimate is, say, between $x = 1$ and $x = 2$, plot new points on the graph, taking for x the values $1\frac{1}{4}$, $1\frac{1}{2}$, $1\frac{3}{4}$, and endeavor to make a closer estimate than before. Estimate the corresponding value of V.

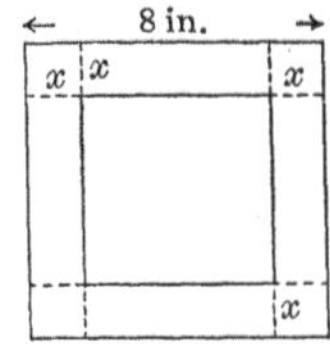

Why were no values of x greater than 4 or less than 0 suggested?

Questions involving *maxima* and *minima*, such as maximum strength, maximum capacity, and minimum cost, are very important both in theory and in practice. The subject is considered at length in the calculus, but analytic geometry affords a valuable method for treating it.

8. The attraction A between the poles of an electromagnet varies inversely as the square of the distance d between the poles; that is, $A = k/d^2$, k being a constant. Taking the case $A = 25/d^2$, plot the graph for all integral values of d from 2 to 10. Estimate from the graph the amount by which A changes between $d = 2$ and $d = 3$; between $d = 8$ and $d = 9$.

9. Certain postal regulations require that the sum of the girth and the length of a parcel to be sent by parcel post shall not exceed 7 ft. Supposing that a manufacturer wishes to ship his goods by parcel post in boxes having square ends and with the girth plus the length equal to 7 ft., study the variation in the capacity of such a box as the dimensions vary.

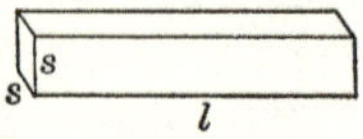

s s l

We evidently have the equations $4s + l = 7$ and $V = s^2l = s^2(7 - 4s)$. Giving to s the values 0, $\frac{1}{2}$, 1, $\frac{3}{2}$, 2, plot the equation $V = s^2(7 - 4s)$, and estimate from the graph the value of s which gives the maximum value of V. Try to improve the accuracy of this estimate, if possible, by closer plotting near the estimated value of s.

10. Consider Ex. 9 for a cylindric parcel of length l inches and radius r inches.

The circumference is $2\pi r$ and the volume is $\pi r^2 l$.

11. A strip of sheet metal 12 in. wide is to be folded along the middle so as to form a gutter. Denoting the width across the top by $2x$, express in terms of x the area A of the cross section of the gutter. Plot the graph of this equation in A and x, and determine approximately the width corresponding to the maximum capacity of the gutter.

12. A rectangular inclosure containing 60 sq. yd. is to be laid off against the wall AB of a house, two end walls of the inclosure being perpendicular to AB, and the other wall being parallel to AB. In terms of the width x of the inclosure, express the total length T of the walls to be built, and plot the graph of this equation in T and x. What value of x makes T the minimum?

13. Find the dimensions of the maximum rectangle that can be inscribed in a circle having a diameter of 16 in.

14. Graphs of Trigonometric Equations. In a trigonometric equation like $y = \sin x$, x is an angle instead of a length; but by letting each unit on the x axis represent a certain angle, we can plot the graph as in the case of an algebraic equation. Taking the values of $\sin x$ from the table on page 285 of the Supplement, we have the following:

$x =$	0°	15°	30°	45°	60°	75°	90°	105°	120°	135°	150°	165°	180°	...
$y =$	0	.26	.50	.71	.87	.97	1	.97	.87	.71	.50	.26	0	...

If we let each unit on the y axis represent 0.20, and each unit on the x axis represent 30°, the graph is as follows:

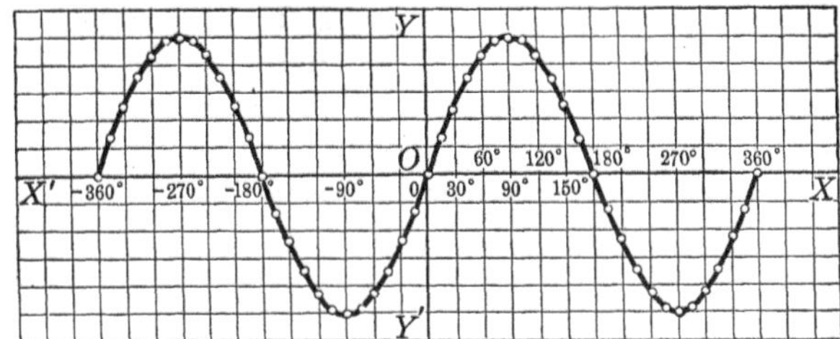

The graph shows clearly many properties of $\sin x$. For example, the maximum value of $\sin x$ is 1 and the minimum value is -1; as x increases from 0° to 90°, $\sin x$ increases from 0 to 1; and so on. The arbitrary selection of 30° as the unit on the x axis does not affect the study of these properties; it affects merely the shape of the curve.

Exercise 4. Trigonometric Graphs

1. From the graph of $y = \sin x$ in § 14 estimate how much y, or $\sin x$, changes when x increases from 0° to 30°; when x increases from 30° to 60°; when x increases from 60° to 90°. Where is the change most rapid? Where is the change the slowest?

2. In Ex. 1 at what values of x is $\sin x$ equal to 1? By how much do these values of x differ?

3. From the graph of $y = \sin x$ in § 14 find between what values of x the value of $\sin x$ is positive, and between what values of x it is negative. At what values of x does $\sin x$ change from positive to negative? from negative to positive?

4. Show how to infer from the graph in § 14 the formulas $\sin(180° - x) = \sin x$ and $\sin(180° + x) = -\sin x$.

5. Draw the graph of $y = \cos x$, from $x = -90°$ to $x = 360°$.

For the equation $y = \cos x$, consider each of the following:

6. Ex. 1. **7.** Ex. 2. **8.** Ex. 3. **9.** Ex. 4.

10. Plot the graphs of $y = \sin x$ and $y = \cos x$ on the same axes; show graphically, as in Ex. 4, that $\sin(90° + x) = \cos x$.

11. The formula for the area of this triangle, as shown in trigonometry, is $A = \frac{1}{2} \cdot 2 \cdot 3 \cdot \sin x$. Plot this equation, showing the variation of A as x increases from 0° to 180°. By looking at the graph, state what value of x makes A greatest.

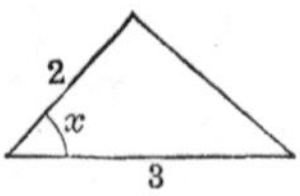

2
x
3

12. From a horizontal plane bullets are fired with a certain velocity, at various angles x with the horizontal. If the distance in yards from the point where a bullet starts to the point where it falls is given by $d = 2750 \sin 2x$, plot this equation, and find the value of x which makes d greatest.

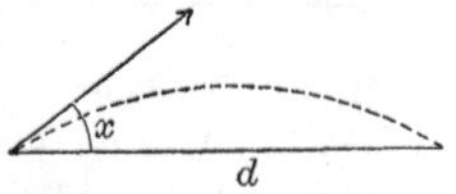

x
d

13. In Ex. 12 estimate from the graph the value of d when $x = 19°$, and estimate the value of x which makes $d = 1050$ yd.

14. If the earth had no atmosphere, the amount H of heat received from the sun upon a unit of surface at P would vary as $\cos x$, where x is the angle between the vertical at P and the direction of the sun from P; that is, $H = k \cos x$. If $x = 27°$ at noon on a certain day, through what values does x vary from sunrise to sunset? Plot the particular case $H = 10 \cos x$ between these values, and find between what values of x the value of H increases most rapidly during the morning.

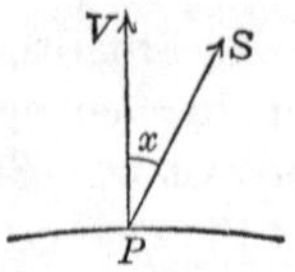

V
S
x
P

CHAPTER II

GEOMETRIC MAGNITUDES

15. Geometric Magnitudes. Although much of our work in geometry relates to magnitudes, the number of different kinds of magnitudes is very small. In plane analytic as well as in synthetic geometry, for example, we consider only those magnitudes which are called to mind by the words *length*, *angle*, and *area*.

In this chapter we shall study the problem of calculating the magnitudes of plane geometry as it arises in our coordinate system.

16. Directed Segments of a Line. Segments measured in opposite directions along a line are said to be positive for one direction and negative for the other.

For example, AB and BA denote the same segment measured in opposite directions. That is, $AB = -BA$, and $BA = -AB$, AB and BA being each the negative of the other.

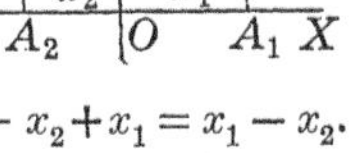

The coordinates of a point are considered as directed segments, the abscissa being always measured from the y axis toward the point, and the ordinate from the x axis toward the point.

That is, in this figure, x_1 is OA_1, not A_1O; but $A_1O = -x_1$. Similarly, $x_2 = OA_2$, but $A_2O = -x_2$. Also, A_2A_1 is not $x_2 + x_1$, for $A_2A_1 = A_2O + OA_1 = -x_2 + x_1 = x_1 - x_2$. These conventions, of little significance in elementary geometry, are of greatest importance in our subsequent work.

17. Distance between Two Points. When the coordinates of two points $P_1(x_1, y_1)$ and $P_2(x_2, y_2)$ are known, the distance d between the points is easily found.

Drawing the coordinates of P_1 and P_2, and drawing P_1Q perpendicular to A_2P_2, we have

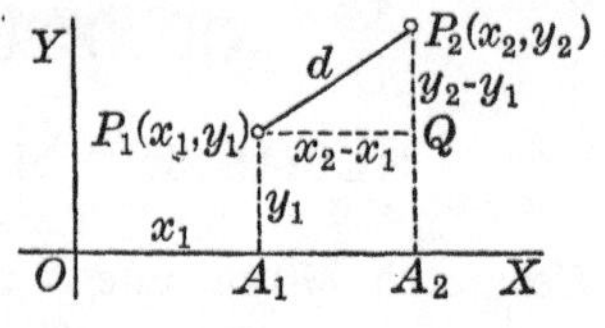

$$P_1Q = OA_2 - OA_1 = x_2 - x_1,$$

$$QP_2 = A_2P_2 - A_2Q = y_2 - y_1;$$

and in the right triangle P_1QP_2

$$\overline{P_1P_2}^2 = \overline{P_1Q}^2 + \overline{QP_2}^2.$$

Therefore, $\mathbf{d = \sqrt{(x_2 - x_1)^2 + (y_2 - y_1)^2}}$.

18. Standard Figure. The figure here shown is used so frequently that we may consider it as a standard. P_1 and P_2 may be any two points in the plane, so that A_1, A_2 and B_1, B_2 may be any two points on the x axis and y axis respectively, determined by P_1 and P_2. But in every case, whether O is to the left of A_1 and A_2, or between them, or to their right,

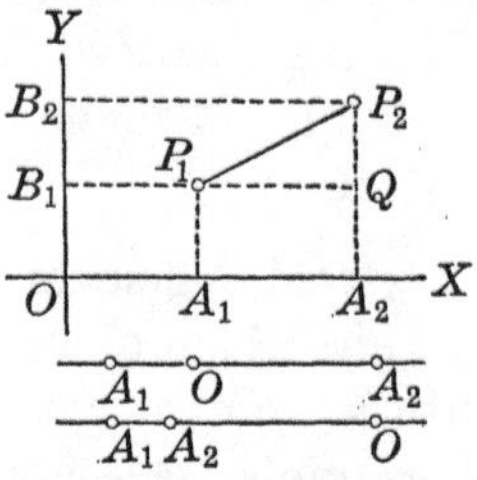

$$A_1A_2 = A_1O + OA_2 = -OA_1 + OA_2 = -x_1 + x_2 = x_2 - x_1,$$

and similarly $B_1B_2 = y_2 - y_1$.

19. Distance in Oblique Coordinates. When the axes are not rectangular, it is the custom to designate the angle XOY by ω. Then, by the law of cosines, we have

$$d^2 = \overline{P_1Q}^2 + \overline{QP_2}^2 - 2\,P_1Q \cdot QP_2 \cos(180° - \omega).$$

That is, since $\cos(180° - \omega) = -\cos\omega$,

$$\mathbf{d = \sqrt{(x_2 - x_1)^2 + (y_2 - y_1)^2 + 2(x_2 - x_1)(y_2 - y_1)\cos\omega}}.$$

Exercise 5. Distance between Two Points

Find the distances between the following pairs of points, drawing the standard figure (§ 18) in each case and specifying the length of each side of the triangle:

1. (3, 3) and (7, 6).
2. $(-6, 2)$ and $(6, -3)$.
3. $(6, 2)$ and $(-2, -4)$.
4. $(-2\frac{1}{2}, 5)$ and $(-8\frac{1}{2}, -3)$.

Find the distances between the following pairs of points:

5. (2, 1) and (5, 1).
6. (6, 0) and (0, 6).
7. $(-6, 0)$ and $(0, -6)$.
8. (0, 0) and (2, 5).
9. (0, 0) and (a, b).
10. (0, 0) and $\left(\frac{1}{2}a, \frac{1}{2}a\sqrt{3}\right)$.

11. Show that the distance of $P(x, y)$ from O is $\sqrt{x^2 + y^2}$.

12. Draw the figure and deduce the formula of § 17 when P_1 is in the second quadrant and P_2 is in the third.

13. Show that (7, 2) and $(1, -6)$ are on a circle whose center is $(4, -2)$, and find the length of the radius.

14. Determine k, given that the distance between (6, 2) and $(3, k)$ is 5. Draw the figure.

Draw the triangles with the following vertices and show by the lengths of their sides that they include equilateral, isosceles, scalene, and right triangles, and that two of them, if regarded as triangles at all, have no area:

15. $(-4, 3)$, $(2, -5)$, $(3, 2)$.
16. $(4, 0)$, $(-4, 0)$, $(0, -4\sqrt{3})$.
17. $(2, 3)$, $(-4, 1)$, $(6, -2)$.
18. $(-6, 8)$, $(6, -8)$, $(8, 6)$.
19. $(5, 1)$, $(2, -2)$, $(8, 4)$.
20. $(-1, -1)$, $(0, 0)$, $(3, 3)$.

21. In this parallelogram $OF_1 = 6$, $OF_2 = 4$, and the angle $F_1OF_2 = 120°$. Using oblique coordinates as in the figure, find OR and F_1F_2.

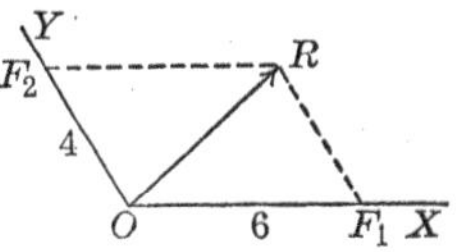

Notice that this is a simple case in the parallelogram of forces.

20. Inclination of a Line. When a straight line crosses the x axis, several angles are formed. The positive angle to the right of the line and above the x axis is called the *angle of inclination* of the line, or simply the *inclination*, and is designated by α.

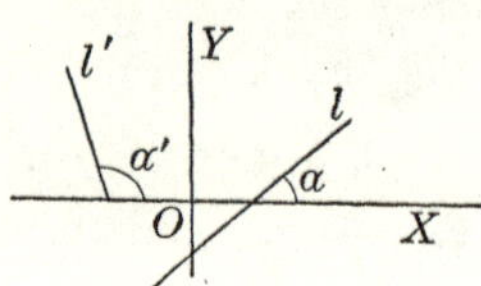

In the above figure the inclination of l is α; that of l' is α'. In analytic geometry an angle is positive if it is generated by the counterclockwise turning of a line about the vertex.

The inclination of a line has numerous important applications to the study of moving bodies, involving such questions as relate to force, work, and velocity.

If the coordinates of two points are given, the inclination of the line determined by these points is easily found.

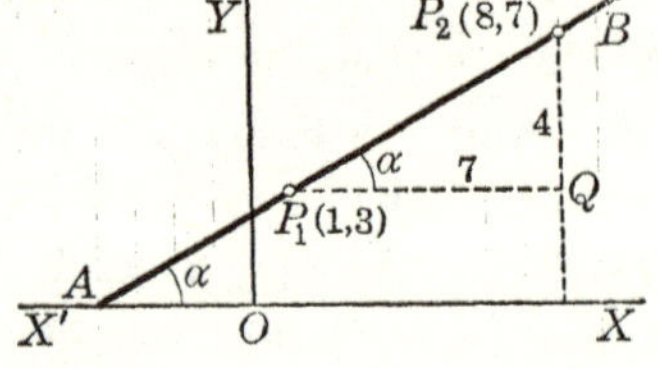

For example, in the case of $P_1(1, 3)$ and $P_2(8, 7)$ we find that $P_1Q = 8 - 1 = 7$, and that $QP_2 = 7 - 3 = 4$; whence $\tan\alpha = \frac{4}{7}$.

21. Slope of a Line. The tangent of the angle of inclination is called the *slope* of the line and is denoted by m. The formula for the slope of the line through two points $P_1(x_1, y_1)$ and $P_2(x_2, y_2)$ is

$$m = \tan\alpha = \frac{y_2 - y_1}{x_2 - x_1}.$$

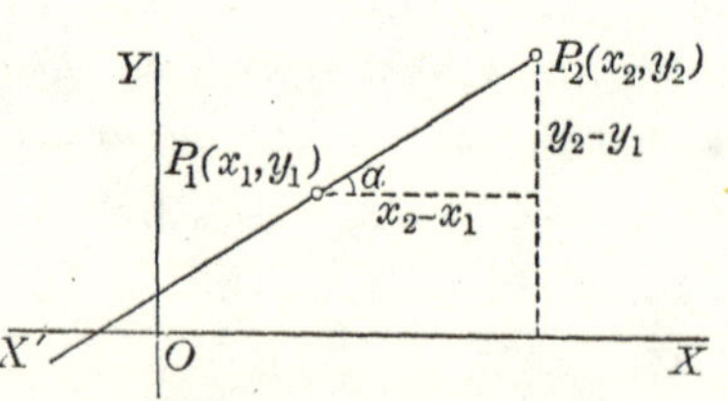

That is, m is equal to the difference between the ordinates divided by the corresponding difference between the abscissas.

If α is between 0° and 90°, m is positive.

If α is between 90° and 180°, m is negative.

If $\alpha = 0°$, $m = \tan 0° = 0$, and the line is parallel to XX'.

If $\alpha = 90°$, the line is perpendicular to XX'.

Exercise 6. Inclinations and Slopes

Find the slope of the line through each of the following pairs of points, and state in each case whether the angle of inclination is an acute, an obtuse, or a right angle:

1. $(2, 3)$ and $(-6, 8)$.

2. $(-1, 4)$ and $(6, -3)$.

3. $(7, -2)$ and $(-3, -2)$.

4. $(-4, -1)$ and $(5, 8)$.

5. $(4, 1)$ and $(4, -4)$.

6. $(3, -\frac{2}{3})$ and $(\frac{4}{3}, \frac{8}{3})$.

7. Show how to construct lines whose slopes are $\frac{1}{2}$, 4, $\frac{2}{3}$, $-\frac{3}{2}$, 1, -1, 0, ∞, and b/a.

8. Draw a square with one side on the x axis, and find the slopes of its diagonals.

9. Draw an equilateral triangle with one side on the x axis and the opposite vertex below the x axis. Find the slope of each side, and the slope of the bisector of each angle.

10. Draw the triangle whose vertices are $(4, -1)$, $(-3, 2)$ and $(-2, 6)$, and find the slope of each side.

11. Show by slopes that the points $(-2, 12)$, $(1, 3)$ and $(4, -6)$ are on one straight line.

12. If $A(-5, -3)$ and $B(3, 7)$ are the ends of a diameter of a circle, show that the center is $(-1, 2)$.

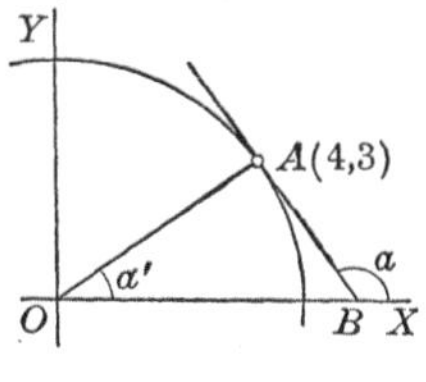

13. Draw the circle with center $(0, 0)$ and radius 5. Show that $A(4, 3)$ is on this circle. Draw AB tangent to the circle at A, and find the slope of AB.

First find $\tan \alpha'$, and then find $\tan \alpha$.

14. If the slope of the line through $(-k, 3)$ and $(5, -k)$ is 1, find the value of k. Plot the points and draw the line.

15. Show that the angle of inclination α of the line through the points $(-2, -4)$ and $(3, 8)$ is twice the angle of inclination α' of the line through the points $(3, -1)$ and $(6, 1)$.

Show that $\tan \alpha = \tan 2\alpha'$.

22. Angle from One Line to Another. When we speak of the angle between AB and CD, we use an ambiguous expression, inasmuch as there are two supplementary angles formed by these lines, namely, θ and ϕ. To avoid this ambiguity we shall speak of the *angle from AB to CD* as the smallest positive angle starting from AB and ending at CD.

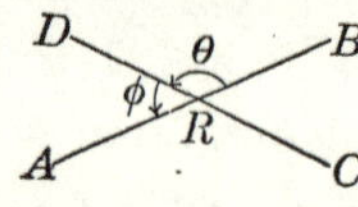

In the figure the angle from AB to CD is θ, either BRD or ARC, but not DRA. The angle from CD to AB is ϕ, either DRA or CRB, but not BRD.

23. Parallel Lines. Since a transversal cutting two parallel lines makes corresponding angles equal, it is evident that parallel lines have the same slope.

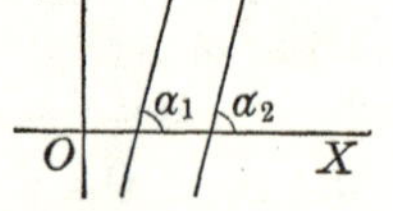

In this figure, $\alpha_1 = \alpha_2$, and hence $m_1 = m_2$, since they are the tangents of equal angles.

The angle from one line to the other is 0° in the case of parallel lines.

24. Perpendicular Lines. If we have two lines, l_1 and l_2, perpendicular to each other, we see by the figure that

$$\alpha_2 = 90° + \alpha_1,$$

and hence we have

$$\tan \alpha_2 = -\cot \alpha_1$$

$$= -\frac{1}{\tan \alpha_1}.$$

Hence $$m_2 = -\frac{1}{m_1},$$

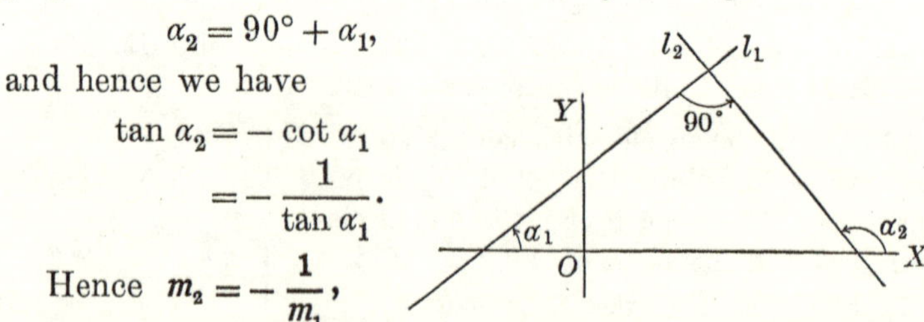

where m_1 is the slope of l_1 and m_2 is the slope of l_2.

That is, *the slope of either of two perpendicular lines is the negative reciprocal of the slope of the other.*

The condition $m_2 = -1/m_1$ is sometimes written $m_1 m_2 = -1$, or $m_1 m_2 + 1 = 0$. This law has many applications in analytic geometry.

25. Angle between Two Lines in Terms of their Slopes. If l_1 and l_2 are two lines with slopes m_1 and m_2 respectively, the angle from l_1 to l_2, which we will designate by θ, is found by first finding $\tan\theta$. In the figure below we see that $\alpha_2 = \theta + \alpha_1$, and hence that $\theta = \alpha_2 - \alpha_1$. Therefore

$$\tan\theta = \tan(\alpha_2 - \alpha_1)$$

$$= \frac{\tan\alpha_2 - \tan\alpha_1}{1 + \tan\alpha_1 \tan\alpha_2};$$

or

$$\boldsymbol{\tan\theta = \frac{m_2 - m_1}{1 + m_1 m_2}}.$$

To find the angle ϕ, from l_2 to l_1, we have

$$\tan\phi = \tan(180° - \theta) = -\tan\theta$$

$$= \frac{m_1 - m_2}{1 + m_1 m_2}.$$

Thus, either arrangement of m_1 and m_2 in the numerator gives one angle or the other; that is, θ or ϕ. If, however, the conditions of a problem call for the angle from the first of two lines to the second, the slope of the second must come first in the numerator.

For example, consider the triangle whose vertices are $A(-5, -3)$, $B(4, -1)$, and $C(3, 4)$. Here the angle B is the angle from BC to BA, or, in the figure, from l_1 to l_2. Then since

$$m_1 = \frac{4 - (-1)}{3 - 4} = -5,$$

and

$$m_2 = \frac{-1 - (-3)}{4 - (-5)} = \frac{2}{9},$$

$$\tan B = \frac{m_2 - m_1}{1 + m_1 m_2} = \frac{\frac{2}{9} - (-5)}{1 + \frac{2}{9}(-5)} = \frac{\frac{47}{9}}{-\frac{1}{9}} = -47.$$

What is indicated with respect to the angle B by the fact that $\tan B$ is negative and also numerically large?

From the figure what is probably the sign of $\tan A$? of $\tan C$?

Exercise 7. Angles

1. The line through $(4, -2)$ and $(2, 3)$ is parallel to the line through $(10, 3)$ and $(12, -2)$.

Hereafter the words "prove that" or "show that" will be omitted in the statements of problems where it is obvious that a proof is required.

2. The line through $(-3, -5)$ and $(7, -2)$ is perpendicular to the line through $(-2, 6)$ and $(1, -4)$.

Find the slope of the line which is perpendicular to the line through each of the following pairs of points:

3. $(2, 1)$ and $(3, -1)$.

4. $(-1, 3)$ and $(4, 2)$.

5. $(-a, 2a)$ and $(1, a)$.

6. $(-6, 0)$ and $(0, 2)$.

7. If a circle is tangent to the line through $(-2, 5)$ and $(4, 3)$, find the slope of the radius to the point of contact.

8. Draw a circle with center $(3, -4)$ and passing through $(5, 2)$. Then draw the tangent at $(5, 2)$ and find its slope.

9. Find the slopes of the sides and also of the altitudes of the triangle whose vertices are $A(3, 2)$, $B(4, -1)$, and $C(-1, -3)$.

10. Show by means of the slopes of the sides that $(0, -1)$, $(3, -4)$, $(2, 1)$, and $(5, -2)$ are the vertices of a rectangle.

11. As in Ex. 10, show that $(2, 1)$, $(5, 4)$, $(4, 7)$, and $(1, 4)$ are the vertices of a parallelogram having a diagonal parallel to the x axis.

12. Draw the triangle whose vertices are $A(2, 1)$, $B(-1, -2)$, and $C(-3, 3)$, and find the tangents of the interior angles of the triangle and also the tangent of the exterior angle at A.

13. If the slopes of l and l' are 3 and m respectively, and the angle from l to l' is $\tan^{-1} 2$, find m.

The symbol $\tan^{-1} 2$ means the angle whose tangent is 2.

14. Find the slope of l, given that the angle from l to the line through $(1, 3)$ and $(2, -2)$ is $45°$; $120°$.

26. Division of a Line Segment. If A and B are the end points of a segment of a line and P is any point on the line, then if P lies between A and B, it divides AB into two parts, AP and PB. Since AP and PB have the same direction, their ratio, designated by $AP:PB$ or by AP/PB, is positive.

If P is either on AB produced or on BA produced, we still speak of it as dividing AB into the two parts AP and PB as before, but in each of these cases the two parts have opposite directions, and hence their ratio is negative.

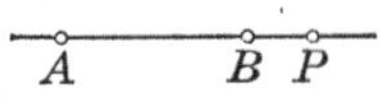

We therefore see that the ratio of the parts is positive when P divides AB internally, and negative when P divides AB externally.

The sum of the two parts is evidently the whole segment; that is, $AP + PB = AB$. Thus, in the last figure, since $AP = -PA$, we have $AP + PB = AP + PA + AB = AB$.

27. Coordinates of the Mid Point of a Line Segment. In the figure below, in which $P_0(x_0, y_0)$ is the mid point of P_1P_2, it is evident from elementary geometry that

$$B_0P_0 = \tfrac{1}{2}(B_1P_1 + B_2P_2),$$

and

$$A_0P_0 = \tfrac{1}{2}(A_1P_1 + A_2P_2),$$

for B_0P_0 and A_0P_0 are medians of the trapezoids $B_1P_1P_2B_2$ and $A_1A_2P_2P_1$. We therefore have

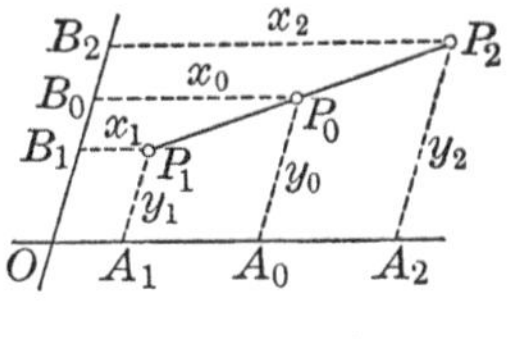

$$x_0 = \frac{x_1 + x_2}{2},$$

and

$$y_0 = \frac{y_1 + y_2}{2}.$$

Although in general we use rectangular coordinates, in this case we have taken oblique axes because the case is so simple. Manifestly, a property which is proved for general axes is true for the special case when the angle O is 90°.

28. Division of a Line Segment in a Given Ratio. If the coordinates of the end points $P_1(x_1, y_1)$, $P_2(x_2, y_2)$ of a line segment P_1P_2 are given, we can find the coordinates of the point $P_0\,(x_0, y_0)$ which divides the segment in a given ratio r.

Since $$r = \frac{P_1P_0}{P_0P_2} = \frac{A_1A_0}{A_0A_2},$$

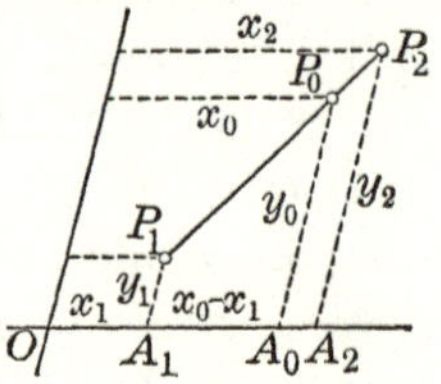

we have $$r = \frac{x_0 - x_1}{x_2 - x_0}.$$

Solving for x_0, $$x_0 = \frac{x_1 + rx_2}{1 + r}.$$

Similarly, $$y_0 = \frac{y_1 + ry_2}{1 + r}.$$

As in § 27, we have taken oblique axes to show the generality of these important formulas.

29. Illustrative Examples. 1. Find the point which divides the line segment from (6, 1) to (– 2, 9) in the ratio 3 : 5.

Since $x_1 = 6$, $x_2 = -2$, and $r = \frac{3}{5}$, we have

$$x_0 = \frac{6 + (-\frac{6}{5})}{1 + \frac{3}{5}} = \frac{\frac{24}{5}}{\frac{8}{5}} = 3.$$

Similarly, $$y_0 = \frac{1 + \frac{27}{5}}{1 + \frac{3}{5}} = \frac{\frac{32}{5}}{\frac{8}{5}} = 4.$$

Hence the required point is (3, 4).

2. Find the point which divides the line segment from (– 4, – 2) to (1, 3) externally in the ratio 8 : 3.

Since $x_1 = -4$, $x_2 = 1$, and $r = -\frac{8}{3}$, we have

$$x_0 = \frac{-4 + (-\frac{8}{3})}{1 + (-\frac{8}{3})} = \frac{-\frac{20}{3}}{-\frac{5}{3}} = 4.$$

Similarly, $$y_0 = \frac{-2 + (-\frac{8}{3}) \cdot 3}{1 + (-\frac{8}{3})} = \frac{-10}{-\frac{5}{3}} = 6.$$

Hence the required point is (4, 6).

Exercise 8. Points of Division

Determine, without writing, the mid point of each of these line segments, the end points being as follows:

1. (7, 4), (3, 2).

2. (6, − 4), (− 2, − 2).

3. (− 3, 1), (2, 7).

4. (4, −1), (− 4, 1).

5. (a, 1), (1, a).

6. (0, 0), (0, $\frac{2}{3}$).

Find the mid point of each of these line segments, the end points being as follows:

7. (7.3, − 4.5), (2.9, 12.3).

8. (− 2.8, 6.4), (− 3.9, 7.2).

9. (14.7, −14.7), (0, 12.2).

10. ($-3\frac{3}{8}$, $-7\frac{5}{8}$), ($2\frac{3}{4}$, $-4\frac{1}{2}$).

Find the point which divides each of these line segments in the ratio stated, drawing the figure in each case:

11. (2, 1) to (3, − 9); 4 : 1.

12. (5, − 2) to (5, 3); 2 : 3.

13. (− 4, 1) to (5, 4); − 5 : 2.

14. (8, 5) to (−13, −2); 4 : 3.

Find the two trisection points of each of these line segments, the end points being as follows:

15. (−1,2), (−10, −1). **16.** (11,6), (2,3). **17.** (7,8), (1, −6).

18. Find the mid points of the sides of the triangle the vertices of which are (7, 2), (−1, 4), and (3, − 6), and draw the figure.

19. In the triangle of Ex. 18 find the point which any median has in common with the other two medians.

By a proposition of elementary geometry the medians are concurrent in a trisection point of each.

20. Given A (4, 7), B (5, 3), and C (6, 9), find the mid points of the sides of the triangle ABC.

21. If one end point of a line segment is A (4, 6) and the mid point is M (5, 2), find the other end point of the segment.

30. Area of a Triangle. Given the coordinates of the vertices of a triangle, the area of the triangle may be found by a method similar to the one used in surveying.

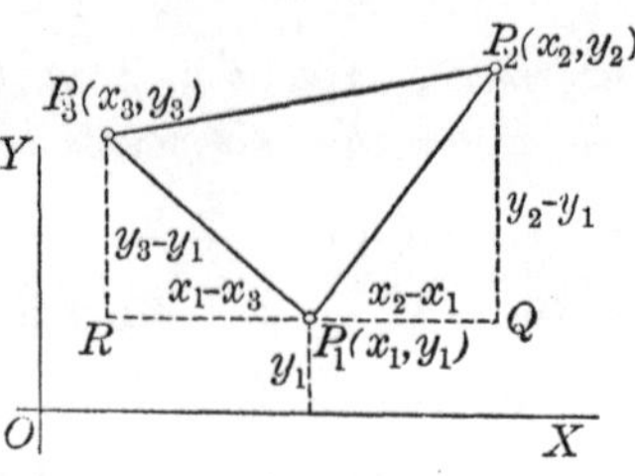

In this figure draw the line RQ through P_1 parallel to OX, and draw perpendiculars from P_3 and P_2 to RQ. Then

$$\triangle P_1P_2P_3 = \text{trapezoid } RQP_2P_3 - \triangle RP_1P_3 - \triangle P_1QP_2.$$

Noting that the altitude of the trapezoid is $x_2 - x_3$, we can easily find the areas of the figures and can show that

$$\triangle P_1P_2P_3 = \tfrac{1}{2}(x_1y_2 - x_2y_1 + x_2y_3 - x_3y_2 + x_3y_1 - x_1y_3).$$

The student who is familiar with determinants will see that this equation may be written in a form much more easily remembered, as here shown.

$$\triangle P_1P_2P_3 = \frac{1}{2}\begin{vmatrix} x_1 & y_1 & 1 \\ x_2 & y_2 & 1 \\ x_3 & y_3 & 1 \end{vmatrix}$$

The area of a polygon may be found by adding the areas of the triangles into which it can be divided.

31. Positive and Negative Areas. Just as we have positive and negative lines and angles, so we have positive and negative areas. When we trace the boundary of a plane figure counterclockwise we consider the area as positive, and when we trace the boundary clockwise we consider the area as negative.

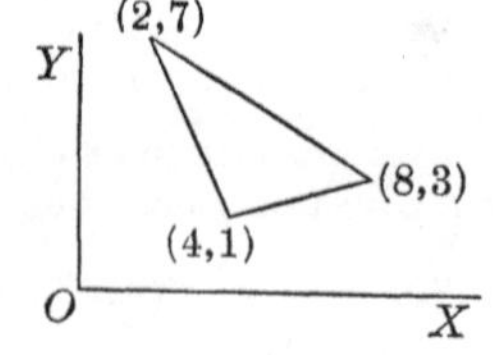

For example, calling the area of this triangle T, if we read the vertices in the order (4, 1), (8, 3), (2, 7), we say that T is positive; but if we read them in the reverse order, we say that T is negative.

Usually, however, we are not concerned with the sign of an area, but merely with its numerical value.

Exercise 9. Areas

Draw on squared paper the following triangles, find by formula the area of each, and roughly check the results by counting the unit squares in each triangle:

1. (3, 3), (−1, −2), (−3, 4).
2. (0, 0), (12, −4), (3, 6).
3. (1, 3), (3, 0), (−4, 3).
4. (−2, −2), (0, 0), (5, 5).

5. Find the area of the quadrilateral (4, 5), (2, −3), (0, 7), (9, 2), and check the result by counting the unit squares.

6. The area of the quadrilateral $P_1(x_1, y_1)$, $P_2(x_2, y_2)$ $P_3(x_3, y_3)$, $P_4(x_4, y_4)$ is

$$\frac{1}{2}(x_1y_2 - x_2y_1 + x_2y_3 - x_3y_2 + x_3y_4 - x_4y_3 + x_4y_1 - x_1y_4).$$

7. Find the point P which divides the median AM of the triangle here shown in the ratio 2 : 1.

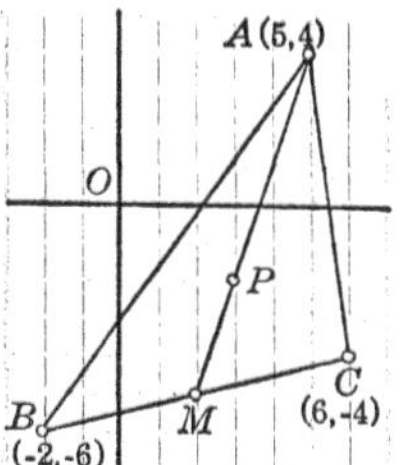

8. In the figure of Ex. 7 join P to A, B, and C, and show that the triangles ABP, BCP, and CAP have the same area.

From elementary geometry we know that P is the common point of the three medians, and that it is often called the *centroid* of the triangle.

In physics and mechanics P is called the *center of area* or *center of mass* of the triangle. Why are these names appropriate?

9. Given the triangle $A(4, -1)$, $B(2, 5)$, $C(-8, 4)$, find the area of the triangle and the length of the altitude from C to AB.

10. Find the lengths of the altitudes of the triangle $A(-4, -3)$, $B(-1, 5)$, $C(3, -4)$.

11. If the area of the triangle $A(6, 1)$, $B(3, 8)$, $C(1, k)$ is 41, find the value of k.

12. By finding the area of ABC, show that the points $A(-1, -14)$, $B(3, -2)$, and $C(4, 1)$ lie on one straight line.

13. Join in order $A(7, 4)$, $B(-1, -2)$, $C(0, 1)$, $D(6, 1)$, and show by Ex. 6 that area $ABCD$ is 0.

Exercise 10. Review

Determine, without writing, the slopes of the lines through the following pairs of points:

1. (3, 4), (5, 7). **3.** (– 2, 2), (2, 4). **5.** (1, a), (a, 1).

2. (3, 3), (4, 1). **4.** (4, – 2), (– 1, 1). **6.** (0, 0), (1, 1).

Determine, without writing, the slopes of lines perpendicular to the lines through the following pairs of points:

7. (2, 4), (4, 2). **9.** (6, 4), (0, 1). **11.** (0, 2), (0, a).

8. (1, 3), (2, 5). **10.** (3, 3), (4, 4). **12.** (4, 0), (a, 0).

13. Find the angle of inclination of the line through (3, 1) and (– 2, – 4).

14. The inclination of the line through (3, 2) and (– 4, 5) is supplementary to that of the line through (0, 1) and (7, 4).

15. If P(5, 9) is on a circle whose center is (1, 6), find the radius of the circle and the slope of the tangent at P.

16. The points (8, 5) and (6, – 3) are equidistant from (3, 2).

17. Given A(2, 1), B(3, – 2), and C(– 4, – 1), show that the angle BAC is a right angle.

18. The line through (a, b) and (c, d) is perpendicular to the line through (b, – a) and (d, – c).

19. Draw the triangle A(4, 6), B(– 2, 2), C(– 4, 6), and show that the line joining the mid points of AB and AC is parallel to BC and equal to half of it.

20. Show that the circle with center (4, 1) and radius 10 passes through (– 2, 9), (10, – 7), (12, – 5). Draw the figure.

21. If the circle of Ex. 20 also passes through (– 4, k), find the value of k.

22. Through (0, 0) draw the circle cutting off the lengths a and b on the axes, and state the coordinates of the center and the length of the radius.

23. If $(3, k)$ and $(k, -1)$ are equidistant from $(4, 2)$, find the value of k.

24. If $(3, k)$ and $(4, -3)$ are equidistant from $(-5, 1)$, find the value of k.

25. If (a, b) is equidistant from $(4, -3)$ and $(2, 1)$, and is also equidistant from $(6, 1)$ and $(-4, -5)$, find the values of a and b and locate the point (a, b).

26. The point $(4, 13)$ is the point of trisection nearer the end point $(3, 8)$ of a segment. Find the other end point.

27. The line AB is produced to C, making BC equal to $\frac{1}{2}AB$. If A and B are $(5, 6)$ and $(7, 2)$, find C.

28. The line AB is produced to C, making $AB:BC = 4:7$. If A and B are $(5, 4)$ and $(6, -9)$, find C.

29. If three of the vertices of a parallelogram are $(1, 2)$, $(-5, -3)$, and $(7, -6)$, find the fourth vertex.

30. Derive the formulas for the coordinates of the mid point of a line as a special case of the formulas of § 28, using the proper special value of r for this purpose.

Plot the points $A(-4, 3)$, $B(2, 6)$, $C(7, -3)$, and $D(-3, -8)$ and show that:

31. These points are the vertices of a trapezoid.

32. The line joining the mid points of AD and BC is parallel to AB and to CD and is equal to half their sum.

33. The line joining the mid points of AD and BC is parallel to AB and to CD and is equal to half their difference.

34. The points $P_1(x_1, y_1)$, $P_2(x_2, y_2)$, and $P_3(x_3, y_3)$ are the vertices of any triangle. By finding the points which divide the medians from P_1, P_2, and P_3 in the ratio $2:1$, show that the three medians meet in a point.

35. The points $A(2, -4)$, $B(5, 2)$, $P(3, -2)$ are all on the same line. Find the point dividing AB externally in the same ratio in which P divides AB internally.

36. Given the four points A (6, 11), $B(-4, -9)$, $C(11, -4)$, and $D(-9, 6)$, show that AB and CD are perpendicular diameters of a circle.

37. Draw the circle with center $C(6, 6)$ and radius 6, and find the length of the secant from (0, 0) through C.

Two circles pass through A (6, 8), and their centers are C (2, 5) and $C'(11, -4)$. Draw the figure and show that:

38. $B(-1, 1)$ is the other common point of the circles.

39. The common chord is perpendicular to CC'.

40. The mid point of AB divides CC' in the ratio 1:17.

Draw the circle with center $C(-1, -3)$ and radius 5, and answer the following:

41. Are $P_1(2, 1)$ and $P_2(-4, 1)$ on the circle?

42. In Ex. 41 find the angle from CP_1 to CP_2.

43. Find the acute angle between the tangents at P_1 and P_2.

Given that P_0 is on the line through P_1P_2 and that r is the ratio P_1P_0/P_0P_2, draw the figure and study the values of r for the following cases:

44. P_0 is between P_1 and P_2. Show that $r > 0$.

45. P_0 is on P_1P_2 produced. Show that $r < -1$.

46. P_0 is on P_2P_1 produced. Show that $-1 < r < 0$.

47. Given the point A (1, 1), find the point B such that the length of AB is 5 and the abscissa of the mid point of AB is 3.

48. In a triangle ABC, if A is the point (4, −1), if the mid point of AB is M (3, 2), and if the medians of the triangle meet at P (4, 2), find C.

49. Find the point Q which is equidistant from the coordinate axes and is also equidistant from the points A (4, 0) and B (−2, 1).

50. Given the points $A(2, 3)$ and $B(8, 4)$, find the points which divide AB in extreme and mean ratio, both internally and externally. Draw the figure and locate the points of division.

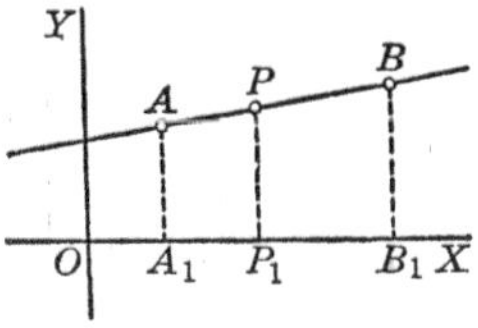

By elementary geometry, P divides AB in extreme and mean ratio if

$$AP : PB = PB : AB.$$

The abscissa of $P(a, b)$ is found by noting that $A_1P_1 : P_1B_1 = P_1B_1 : A_1B_1$, and that $A_1P_1 = a - 2$, $P_1B_1 = 8 - a$, and $A_1B_1 = 8 - 2$. Similarly the ordinate b can be found. The two roots of the quadratic equation in a are the abscissas for both the internal and external points of division, only the former being shown in this figure.

51. Given the points $A(-3, -8)$ and $B(3, -2)$, find the points $P_1(1, b)$ and $P_2(c, d)$ which divide AB internally and externally in the same ratio.

52. The vertices of a triangle are $A(-6, -2)$, $B(6, -5)$, and $C(2, -8)$, and the bisector of angle C meets AB at D. Find D.

Recall the fact that the bisector of an interior angle divides the opposite side into parts proportional to the other two sides.

53. In Ex. 52 find the point in which the bisector of the exterior angle at C meets AB.

Given $A(1, -3)$ and $B(4, 1)$, find all points P meeting the following conditions, drawing the complete figure in each case:

54. $P(3, k)$, given that AP is perpendicular to PB.

55. $P(5, k)$, given that AP is perpendicular to PB.

Explain any peculiar feature that is found in connection with Ex. 55.

56. $P(2, k)$, given that angle $APB = 45°$.

57. $P(k, 10)$, given that angle $APB = 45°$.

Explain any peculiar feature that is found in connection with Ex. 57.

58. $P(a, a)$, given that angle $APB = 45°$.

59. $P(a, b)$, given that $AP = PB$ and the slope of OP is 1.

60. $P(a, b)$, given that $OP = 5$ and the slope of OP is 2.

61. Given that $AP = PB$, AP is perpendicular to PB, A is $(2, 3)$, and B is $(-3, -2)$, find the coordinates of P.

62. The mid point of the hypotenuse of a right triangle is equidistant from the three vertices.

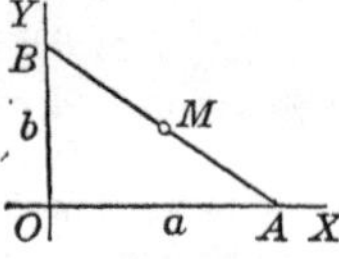

This theorem, familiar from elementary geometry, makes no mention of axes. We, therefore, choose the most convenient axes, which, in this case, are those which lie along OA and OB. From the coordinates of $A(a, 0)$ and of $B(0, b)$, we can find those of M and the three distances referred to in the exercise.

63. In any triangle ABC the line joining the mid points of AB and AC is parallel to BC and is equal to $\frac{1}{2}BC$.

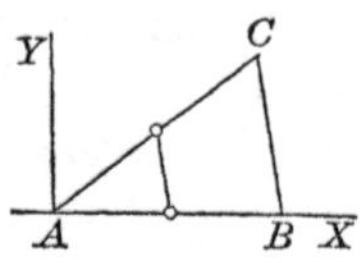

The triangle may be regarded as given by the coordinates of the vertices.

64. Using oblique axes, prove from this figure that the diagonals of a parallelogram bisect each other.

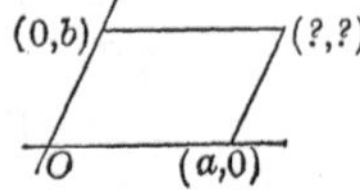

In proving this familiar theorem by analytic geometry it is convenient to use oblique axes as shown by the figure. Show that the diagonals have the same mid point.

65. The lines joining in succession the mid points of the sides of any quadrilateral form a parallelogram.

66. Using rectangular axes as shown in this figure, prove that the diagonals of a parallelogram bisect each other.

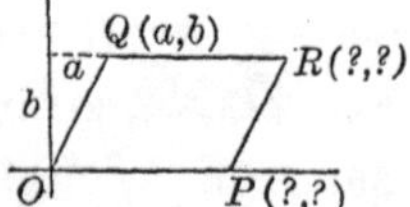

67. The diagonals of a rhombus are perpendicular to each other.

Since we have not yet studied the slope of a line with respect to oblique axes, we choose rectangular axes for this case. We may use the figure of Ex. 66, making OP equal to OQ. Assuming that the coordinates of Q are a and b, what are the coordinates of P and R?

68. The vertices of a triangle are $A(0, 0)$, $B(4, 8)$, and $C(6, -4)$. If M divides AB in the ratio $3:1$, and P is a point on AC such that the area of the triangle APM is half the area of ACB, in what ratio does P divide AC?

CHAPTER III

LOCI AND THEIR EQUATIONS

32. Locus and its Equation. In Chapter I we represented certain geometric loci algebraically by means of equations.

For example, we saw that the coordinates of the point $P(x, y)$, as this point moves on a circle of radius r and with center at the origin, satisfies the equation $x^2 + y^2 = r^2$. This is, therefore, the equation of such a circle.

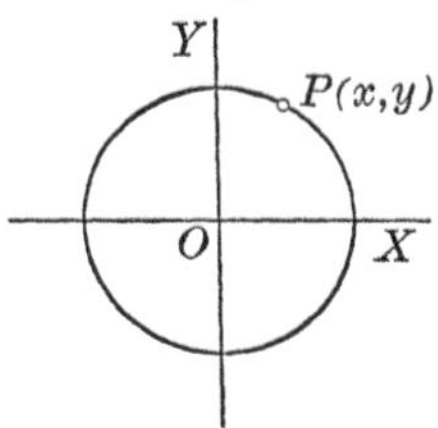

In this connection two fundamental problems were stated on page 9:

1. *Given the equation of a graph, to draw the graph.*
2. *Given a graph, to find its equation.*

It is the purpose of this chapter to consider some of the simpler methods of dealing with these two problems.

We have already described on pages 10–14 a method of dealing with the first of the problems; but this method, when used alone, is often very laborious.

33. Geometric Locus. In elementary geometry the locus of a point is often defined as the path through which the point moves in accordance with certain given geometric conditions. It is quite as satisfactory, however, to disregard the idea of motion and to define a locus as a set of points which satisfy certain given conditions.

Thus, when we say that a locus of a point is given, we mean that the conditions which are satisfied by the points of a certain set are given, or that we know the conditions under which the point moves.

34. Equation of the Line through Two Points. The method of finding the equation of the straight line through two given points is easily understood from the special case of the line through the points $A(2, -3)$ and $B(4, 2)$.

If $P(x, y)$ is any point on the line, it is evident that the slope of AP is the same as the slope of AB. Hence, by § 21,

$$\frac{y+3}{x-2}=\frac{5}{2};$$

that is, $$2y-5x+16=0.$$

If $P(x, y)$ is not on the line, the slope of AP is not the same as the slope of AB, and this equation is not true.

This equation, being true for all points $P(x, y)$ on the line and for no other points, is the equation of the line.

It should be observed that the equation is true for all points on the unlimited straight line through A and B and not merely for all points on the segment AB.

35. Equation of a Line Parallel to an Axis. Let the straight line AB be parallel to the y axis and 5 units to the right. As the point $P(x, y)$ moves along AB, y varies, but x is always equal to 5. The equation $x=5$, being true for all points on AB and for no other points, is the equation of the line.

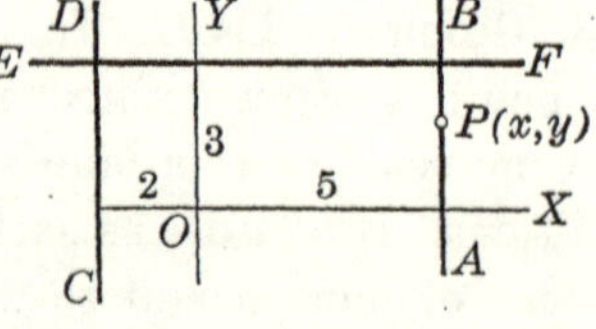

Similarly, the equation of the line CD, parallel to the y axis and 2 units to the left, is $x=-2$.

Similarly, the equation of the line EF, parallel to the x axis and 3 units above, is $y=3$.

In general, the equation of the line parallel to the y axis and a units from it is $x=a$, and the equation of the line parallel to the x axis and b units from it is $y=b$.

Exercise 11. Equations of Straight Lines

1. Find the equation of the line which passes through $(4, -1)$ and $(-2, 2)$.

2. Given the triangle with vertices $(4, -2)$, $(-2, -3)$, and $(2, 4)$, find the equations of the sides.

3. In Ex. 2 find the equations of the medians, and show that these equations have a common solution. What inference may be drawn from the fact of the common solution?

4. Find the equation of the line through $(3, 1)$ and $(-2, -9)$, and show that the line passes through $C(5, 5)$.

Show that the coordinates of C satisfy the equation of the line.

5. Find the equation of the line through $(3, 3)$ and $(-2, 4)$ and that of the line through $(1, 1)$ and $(5, -1)$, and show that $(-7, 5)$ is on both lines.

Find the equations of the lines through the following points:

6. $(3, 1)$ and $(-2, a)$.

7. $(2, 4a)$ and $(1, 2a)$.

8. $(-1, -2)$ and $(-2, -4)$.

9. (a, b) and (c, d).

10. Find the equation of the line which passes through $A(3, -2)$ and has the slope $\frac{2}{3}$.

The condition under which $P(x, y)$ moves is that the slope of AP is $\frac{2}{3}$.

Find the equations of the lines through the following points and parallel to the x axis:

11. $(4, -2)$. **12.** $(4, 7)$. **13.** $(a, 0)$. **14.** $(-a, -b)$.

Find the equations of the lines through the following points and parallel to the y axis:

15. $(2, -1)$. **16.** $(6, 6)$. **17.** $(0, m)$. **18.** $(-p, -q)$.

19. The equation of the x axis is $y = 0$.

20. Find the equation of the y axis.

21. The points $(1, 1)$, $(2, 0)$, $(0, 2)$ lie on a straight line.

36. Equation of a Circle. The general equation of the circle with a given center and a given radius is easily deduced after considering a special case.

Let $C(2, 3)$ be the center and 7 the radius of a given circle. Then if $P(x, y)$ is any point on the circle, we have

$$CP = 7,$$

whence $$(x-2)^2+(y-3)^2=49,$$

and $$x^2+y^2-4x-6y=36,$$

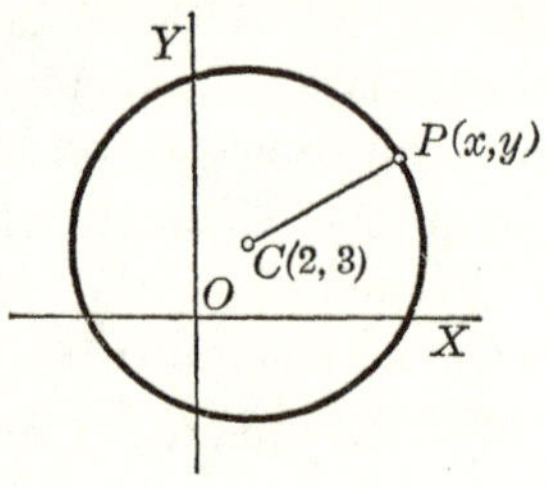

which is true for all points (x, y) on the circle and for no other points.

In general, if $C(a, b)$ is the center and r the radius of any circle in the plane, and if $P(x, y)$ is any point on the circle, the equation of the circle may be found from the fact that

$$CP = r.$$

Hence $$\sqrt{(x-a)^2+(y-b)^2}=r; \qquad \S 17$$

that is, $$\mathbf{(x-a)^2+(y-b)^2=r^2},$$

or $$\mathbf{x^2+y^2-2ax-2by+c=0},$$

where c stands for $a^2+b^2-r^2$.

The form of this equation should be kept in mind so as to be instantly recognized in the subsequent work. The equation has terms in x^2, y^2, x, y, and a constant term, and the coefficients of x^2 and y^2 are equal and have the same sign. In special cases any or all of the numbers a, b, and c may be 0.

For example, $x^2+y^2-6x+8y=11$ represents the circle with center $(3, -4)$ and radius 6, since the equation may be written

$$x^2-6x+9+y^2+8y+16=11+9+16,$$

or $$(x-3)^2+(y+4)^2=36.$$

In particular, the equation of the circle with center $O(0, 0)$ and radius r is

$$\mathbf{x^2+y^2=r^2}.$$

We shall study circles and their equations further in Chapter V.

Exercise 12. Equations of Circles

Draw the following circles and find the equation of each:

1. Center (4, 3), tangent to the x axis.
2. Center (– 5, 0), tangent to the y axis.
3. Through (0, – 8), tangent to the x axis at the origin.
4. Tangent to OY and to the lines $y = 2$ and $y = 6$.
5. Tangent to the lines $x = -1$, $x = 5$, and $y = -2$.
6. Through $A(-6, 0)$, $B(0, -8)$, and $O(0, 0)$.

Notice that AB is a diameter of the circle.

7. Find the equation of the circle with center (2, – 2) and radius 13; draw the circle; locate the points (– 3, 10), (14, 3), (3, 10), (2, 11), and (2, – 15); and determine through which of these points the circle passes.

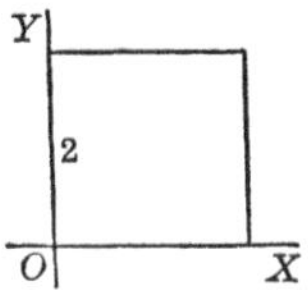

8. Find the equation of the circle inscribed in the square shown at the right.

9. Find the equation of the circle circumscribed about the square shown in the figure above.

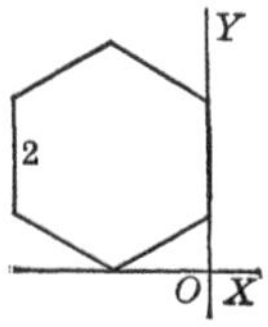

10. Find the equation of the circle inscribed in the regular hexagon here shown.

11. Find the equation of the circle circumscribed about the regular hexagon in Ex. 10.

Find the center and the radius of each of the circles represented by the following equations, and draw each circle:

12. $x^2 + y^2 - 2x - 6y = 15$.
13. $x^2 + y^2 + 6x + 3y = 1$.
14. $2x^2 + 2y^2 - 4x + 10y = 21$.
15. $x^2 + y^2 = 6x$.
16. $x^2 + y^2 + 2x - 4y = 20$.
17. $x^2 + y^2 = 6y + 16$.

18. On the circle $x^2 + y^2 - 6x + 2y = 7$ find each point whose ordinate is 3, and find each point in which the circle cuts the x axis.

37. Equations of Other Loci. The method set forth in §§ 34–36 enables us to find the equations of many other important curves. For example, consider the case of the locus of a point which moves so that its distance from the point (4, 0) is equal to its distance from the y axis.

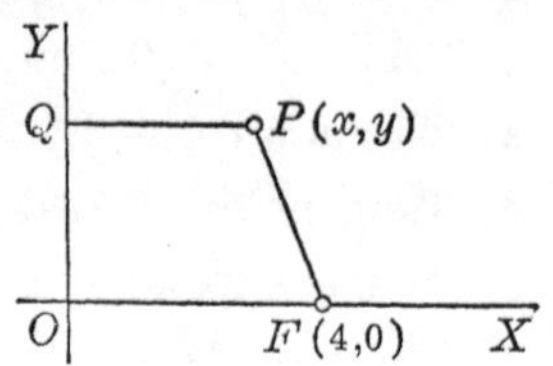

This being a locus not thus far considered, we shall not attempt to draw it in advance, but shall first take any point $P(x, y)$ which satisfies the given condition

$$FP = QP.$$

Expressing FP in terms of x and y (§ 17), and remembering that QP is x, the abscissa of P, we have

$$\sqrt{(x-4)^2+y^2}=x,$$

whence $$y^2-8x+16=0,$$

which is the required equation.

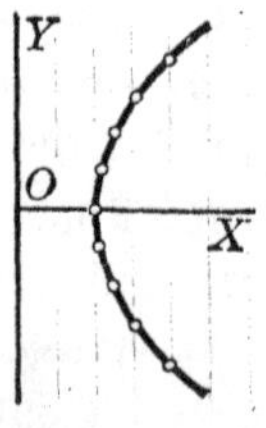

Assigning values to y, we have:

When $y=$	0	± 1	± 2	± 3	± 4
then $x=$	2	$2\frac{1}{8}$	$2\frac{1}{2}$	$3\frac{1}{8}$	4

Plotting these points and connecting them by a smooth curve, we have the locus here shown.

38. General Directions. The above method of finding the equation of a given locus may be stated briefly as follows:

1. *Draw the axes, and denote by $P(x, y)$ any point which satisfies the given geometric condition.*

2. *Write the given condition in the form of an equation.*

3. *Write this equation in terms of x and y, and simplify the result.*

Exercise 13. Equations of Other Loci

Find the equation of the locus of the point P and draw the locus, given that P moves subject to the following conditions:

1. The sum of the squares of its distances from (3, 0) and (– 3, 0) is 68.

2. The sum of the squares of its distances from (2, –1) and (– 4, 5) is 86.

3. Its distance from (0, 6) is equal to its distance from OX.

The resulting equation should be cleared of radicals.

4. Its distance from (8, 2) is twice its distance from (4, 1).

5. Its distance from (3, 0) is 1 less than its distance from the axis OY.

6. Its distance from (–1, 0) is equal to its distance from the line $x = 5$.

7. Its distance from OX exceeds its distance from the point $(-\frac{1}{2}, 1)$ by $\frac{1}{2}$.

8. The angle $APB = 45°$, where A is the point (4, –1) and B is the point (– 2, 3).

9. The angle $APB = 135°$, A and B being as in Ex. 8.

10. Tan $APB = \frac{1}{2}$, where A is (5, 3) and B is (–1, 7).

11. Area PAB = area PQR, where A is (2, 5), B is (2, 1), Q is (3, 0), and R is (5, 3).

12. Its distance from A (4, 0) is equal to 0.8 of its distance from the line $x = 6.25$.

13. The sum of its distances from A (4, 0) and B (– 4, 0) is 10.

14. Its distance from A (5, 0) is $\frac{5}{4}$ its distance from the line $x = 3.2$.

15. The difference of its distances from (5, 0) and (– 5, 0) is 8.

16. The product of its distances from the axes is 10.

17. The sum of the squares of its distances from (0, 0), (4, 0), and (2, 3) is 41.

Find the equation of the locus of the point P under the conditions of Exs. 18–21:

18. The distance of P from the line $y = 3$ is equal to the distance of P from the point $(0, -3)$.

19. The sum of the distances of P from the sides of a square is constant, the axes passing through the center of the square and being parallel to the sides.

20. The sum of the squares of the distances of P from the sides of a square is constant, the axes being as in Ex. 19.

21. A moving ordinate of the circle $x^2 + y^2 = 36$ is always bisected by P.

22. The rod AB in the figure is 4 ft. long, has a knob (P) 1 ft. from A, and slides with its ends on OX and OY respectively. Find the equation of the locus of P.

From the conditions of the problem P moves so that $OK : KA = 3 : 1$.

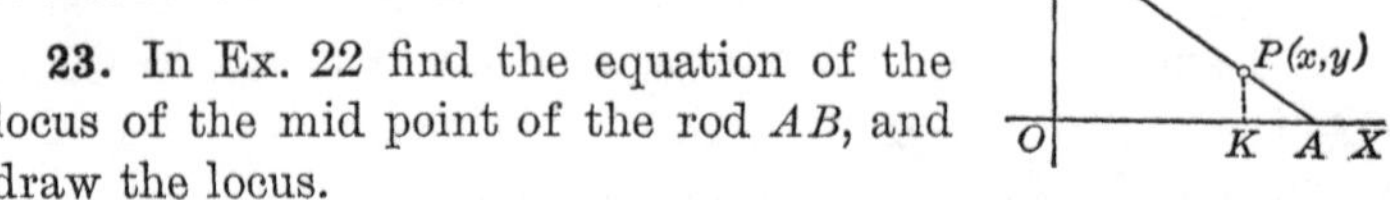

23. In Ex. 22 find the equation of the locus of the mid point of the rod AB, and draw the locus.

24. If every point P of the plane is attracted towards $O(0, 0)$ with a force equal to $10/\overline{OP}^2$, and towards $A(12, 0)$ with a force equal to $5/\overline{AP}^2$, find the equation of the locus of all points P which are equally attracted towards O and A, and draw this locus.

25. If a certain spiral spring 6 in. long, attached at $A(8, 0)$, is extended e inches to P, the pull at P is $5e$ pounds. A 3-inch spring of like strength being attached at $B(-2, 0)$, it is required to find the equation of the locus of all points in the plane at which the pull from A is twice that from B.

The pull along PA is not five times the number of inches in PA, but only five times the excess of this length over 6 in., the original length of spring. It is assumed that the coordinates are measured in inches.

Similarly, the pull along PB is five times the excess of the length of PB over 3 in.

39. Graph of an Equation. We shall now consider the other fundamental problem of analytic geometry, given on pages 9 and 33: *Given an equation, to draw its graph.*

Although simple methods for solving this problem have already been given, certain other suggestions concerning it will be helpful.

40. Equations already Considered. We have already considered a few equations of the first degree, and have inferred that $Ax + By + C = 0$ represents a straight line.

We have also shown (§ 36) that the equation

$$x^2 + y^2 - 2ax - 2by + c = 0$$

represents the circle of center (a, b) and radius $\sqrt{a^2 + b^2 - c}$.

41. Other Equations. As shown on page 10, in the case of other equations any desired number of points of the graph may be found by assigning values to either of the coordinates, x or y, and computing the corresponding values of the other.

The closer together the points of a graph are taken, the more trustworthy is our conception of the form of the graph. If too few points are taken, there is danger of being misled.

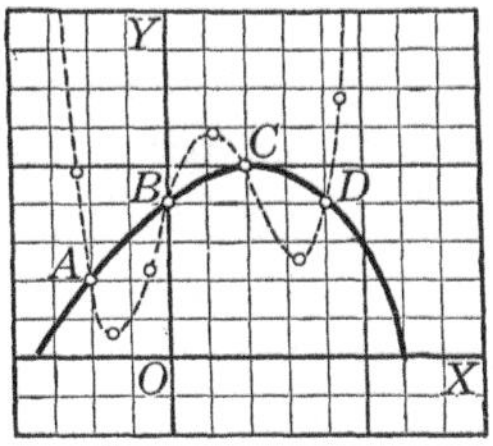

For example, consider the equation

$$12y = 12x^4 - 25x^3 - 15x^2 + 34x + 24.$$

When $x =$	-1	0	1	2
then $y =$	1	2	2.5	2

If we locate only A, B, C, D in the figure above, the heavy line might be thought to be the graph; but if we add the six points in the table below, the graph is seen to be more like the dotted line.

When $x =$	-1.2	-0.7	-0.3	0.6	1.7	2.2
then $y =$	2.47	0.36	1.10	2.93	1.32	3.43

42. Examination of the Equation. Since the tentative process of locating the points of the graph of an equation is often laborious, we seek to simplify it in difficult cases by an examination of the equation itself. For example, taking the equation

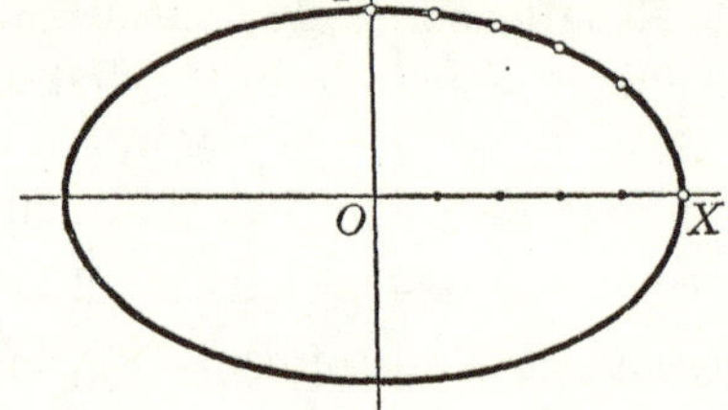

$$9x^2 + 25y^2 = 225,$$

we have

$$y = \pm \tfrac{3}{5}\sqrt{25 - x^2},$$

from which the following facts are evident:

1. The variable x may have any value from -5 to 5 inclusive; but if $x < -5$ or if $x > 5$, y is imaginary.

2. To every value of x there correspond two values of y, numerically equal but of opposite sign. It is evident that the graph is symmetric with respect to the x axis.

3. Two values of x, equal numerically but of opposite sign, give values of y that are equal. It is therefore evident that the graph is symmetric with respect to the y axis.

For example, if $x = 2$ or if $x = -2$, we have $y = \pm \frac{3}{5}\sqrt{21}$. We may therefore locate points to the right of the y axis and merely duplicate them to the left.

4. As x increases numerically from 0 to 5, y decreases numerically from 3 to 0.

From this examination of the equation we see that it is necessary to locate the points merely in the first quadrant, draw the graph in that quadrant, and duplicate this in the other quadrants. For the first quadrant we have:

When $x =$	0	1	2	3	4	5
then $y =$	3	2.9	2.7	2.4	1.8	0

43. Symmetry. The notion of symmetry is familiar from elementary geometry, and therefore the term was freely employed in § 42. We shall, however, use the word frequently as we proceed, and hence it is necessary to define it and to consider certain properties of symmetry.

Two points are said to be *symmetric with respect to a line* if the line is the perpendicular bisector of the line segment which joins the two points.

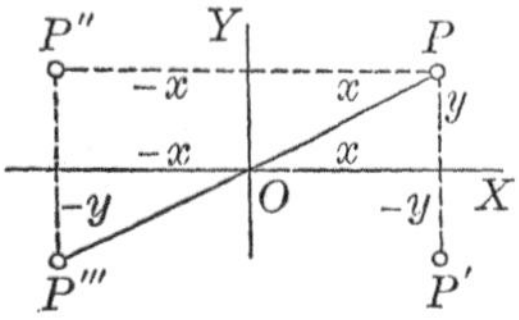

For example, $P(x, y)$ and $P'(x, -y)$ in this figure are symmetric with respect to OX.

Two points are said to be *symmetric with respect to a point* if this point bisects the line segment which joins the two points.

Thus, $P(x, y)$ and $P'''(-x, -y)$ in the figure above are symmetric with respect to the origin O.

A graph is said to be *symmetric with respect to a line* if all points on the graph occur in pairs symmetric with respect to the line; and a graph is said to be *symmetric with respect to a point* if all points on the graph occur in pairs symmetric with respect to the point.

The following theorems are easily deduced:

1. *If an equation remains unchanged when y is replaced by $-y$, the graph of the equation is symmetric with respect to the x axis.*

For if (x, y) is any point on the graph, so is $(x, -y)$.

2. *If an equation remains unchanged when x is replaced by $-x$, the graph of the equation is symmetric with respect to the y axis.*

3. *If an equation remains unchanged when x and y are replaced by $-x$ and $-y$ respectively, the graph of the equation is symmetric with respect to the origin.*

44. Interval. On page 42, in studying the graph of the equation $9x^2 + 25y^2 = 225$, we suggested considering the values of x in three sets of numbers: (1) from $-\infty$ to -5, (2) from -5 to 5, and (3) from 5 to ∞.

The set of all real numbers from one number a to a larger number b is called the *interval* from a to b.

The notation $a \leqq x \leqq b$ means that x is equal to or greater than a, and is equal to or less than b; that is, it means that x may have any value in the interval from a to b, inclusive of a and b.

The notation $a < x < b$ means that x may have any value in the interval from a to b, exclusive of a and b.

In examining a quadratic equation for the purpose of determining the intervals of values of x for which y is real, we are often aided by solving the equation for y in terms of x, factoring expressions under radicals if possible.

For example, from the equation

$$y^2 - 2x^2 + 12x - 10 = 0$$

we have

$$y = \pm\sqrt{2(x-1)(x-5)}.$$

If $-\infty \leqq x < 1$, both $x - 1$ and $x - 5$ are negative, and y is real.
If $1 < x < 5$, then $x - 1$ is positive, $x - 5$ negative, and y imaginary.
If $5 < x \leqq \infty$, both $x - 1$ and $x - 5$ are positive, and y is real.

45. Intercepts. The abscissas of the points in which a graph cuts the x axis are called the *x intercepts* of the graph.

The ordinates of the points in which a graph cuts the y axis are called the *y intercepts* of the graph.

Since the ordinate of every point on the x axis is 0 and the abscissa of every point on the y axis is 0,

To calculate the x intercepts of the graph of an equation, put 0 for y and solve for x.

To calculate the y intercepts of the graph of an equation, put 0 for x and solve for y.

Exercise 14. Graphs of Equations

Examine the following equations as to intercepts, symmetry, the x intervals for which y is real, and the x intervals for which y is imaginary, and draw the graphs:

1. $y^2 - 2x^2 + 12x = 10.$
2. $y^2 + 8x + 16 = 0.$
3. $8y^2 - x^3 = 0.$
4. $4y^2 - 9x^2 = 27x - 90.$
5. $4x^2 + 9y^2 - 36 = 0.$
6. $4x^2 - 9y^2 - 36 = 0.$
7. $9x^2 + 4y^2 - 36 = 0.$
8. $3x = 6 - 4y^2.$
9. $y^2 + 9x^2 - 9 = 0.$
10. $x^2 + 4y^2 - 6x = 16.$
11. $3x^2 - 56x + 240 = y^2.$
12. $8x^2 - y^2 - 8 = 0.$
13. $x^2 + 9y^2 = 9.$
14. $x^2 - y^2 - 8 = 0.$
15. $y^2 - x^2 - 8 = 0.$
16. $4x^2 + 9y^2 = 72.$
17. $4x^2 + 3y^2 = 16x - 12.$
18. $(y - 5)^2 = x^2 - x - 12.$
19. $y^2 = (x - 1)(x - 3)(x - 4).$
20. $y^2 = 2x^3 + 10x^2 - 28x.$
21. $9y^2 = 4x(8 - 2x - x^2).$
22. $y^2 = (x - 1)(x - 3)^2.$

In Ex. 9 write $y = \pm\sqrt{-9x^2 + 9} = \pm 3\sqrt{-(x+1)(x-1)}$.

In Ex. 11 write $y = \pm\sqrt{3x^2 - 56x + 240} = \pm\sqrt{(3x - 20)(x - 12)}$ $= \pm\sqrt{3(x - \frac{20}{3})(x - 12)}$.

In Ex. 14 write $y = \pm\sqrt{x^2 - 8} = \pm\sqrt{(x + \sqrt{8})(x - \sqrt{8})}$.

Examine the following equations as above, arranging each as a quadratic in y and using the quadratic formula:

23. $x^2 - 2xy + 2y^2 + 6x - 8y + 2 = 0.$

As a quadratic in y we have $2y^2 - 2(x + 4)y + x^2 + 6x + 2 = 0$. Using the quadratic formula as given in the Supplement on page 283,

$$y = \frac{2(x + 4) \pm \sqrt{4(x + 4)^2 - 8(x^2 + 6x + 2)}}{4}$$

$$= \frac{x + 4 \pm \sqrt{-x^2 - 4x + 12}}{2} = \frac{x + 4 \pm \sqrt{-(x + 6)(x - 2)}}{2}.$$

24. $x^2 - 4y^2 + 6x + 8y + 1 = 0.$
25. $x^2 - y^2 - 4x - 6y - 5 = 0.$

Examine each of the following equations with respect to intercepts and symmetry, solve for x in terms of y, determine the y intervals for which x is real or imaginary, assign values to y, and draw the graph:

26. $3x = 6 - 4y^2$.

27. $y^2 - x^2 - 4y = 0$.

28. $(y - x)^2 = 8y - 16$.

29. $4(x^2 + y^2) = y^4$.

30. $3x^2 + 2y^2 - 2y = 12$.

31. $4x^2 + 17y^2 = y^4 + 16$.

In Ex. 29 the point (0, 0) is called an *isolated point* of the graph.

32. A rectangle is inscribed in a circle of radius 12. If one side of the rectangle is $2x$, find the area A of the rectangle in terms of x, plot the graph of the equation in A and x, and estimate the value of x which gives the greatest value of A.

33. In Ex. 32 express the perimeter P in terms of x, draw the graph of the equation in P and x, and estimate the value of x which gives the least value of P.

34. Consider Ex. 32 for a rectangle inscribed in a semicircle.

35. Find by a graph the volume of the greatest cylinder that can be inscribed in a sphere of radius 10 in.

36. In this figure A and B are two towns, AQ is a straight river, BQ is perpendicular to AQ, $AQ = 8$ mi., and $BQ = 5$ mi. A pumping station P is to be built on the river to supply both towns with water. Express in terms of x the total cost C of laying the pipes at \$600 a mile for AP and \$1000 a mile for PB, plot the equation, and estimate the most economical position for the pumping station.

37. Two sources of heat, A and B, are 10 ft. apart, and A gives out twice as much heat as B. If P is on the line AB and is x feet from A and x' feet from B, and receives $50/x^2$ units of heat from A and $25/x'^2$ units of heat from B, express the total number H of units of heat received by P from both A and B, draw the graph of the equation, and find the coolest position for P between A and B.

46. Asymptote. If P is a point on the curve C and PA is perpendicular to the line l, then if $PA \rightarrow 0$ when P moves indefinitely far out on C, the line l is said to be an *asymptote* of the curve.

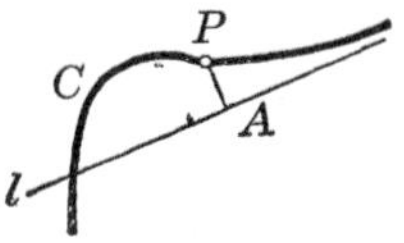

The notation $PA \rightarrow 0$, which was used in the definition above, and for which the notation $PA \doteq 0$ is also sometimes used, is read, "PA approaches zero as a limit." In general, $x \rightarrow a$ means "x approaches a as a limit"; but $x \rightarrow \infty$ means "x increases without limit."

At present we shall consider only those asymptotes which are parallel to the y axis or x axis, calling them respectively *vertical asymptotes* and *horizontal asymptotes*. Methods of dealing with other asymptotes are given in the calculus.

An asymptote is often a valuable guide in drawing a curve.

47. Vertical Asymptote. The advantage of considering the vertical asymptote in certain cases may be seen in examining the equation $xy - 2x - 4y + 10 = 0$. Solving for y, we have

$$y = \frac{2x - 10}{x - 4},$$

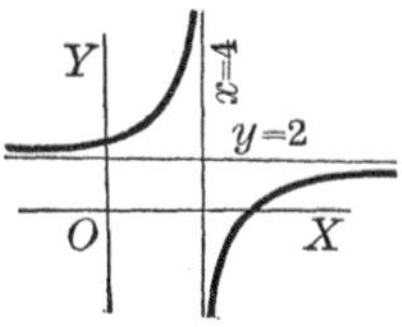

with respect to which the following observations are important:

1. When $x < 4$, let x approach 4 as a limit; $2x - 10$ is negative and approaches -2 as a limit, and $x - 4$ is negative and approaches 0 as a limit. Hence y is positive and increases without limit.

Drawing the vertical line $x = 4$, we see that the graph approaches the line $x = 4$ as an asymptote.

2. When $x > 4$, let x approach 4 as a limit; $2x - 10$ approaches -2 as a limit, and $x - 4$ is positive and approaches 0 as a limit. Hence y is negative and decreases without limit.

We see that as the graph approaches the line $x = 4$ it runs from the left upward toward the line, and runs from the right downward.

48. Horizontal Asymptote. Taking the equation of § 47, we shall now consider the graph as it extends indefinitely to the right; that is, as x increases without limit.

As x increases without limit the numerator $2x-10$ and the denominator $x-4$ both increase without limit and the value of y appears uncertain; but if we divide both terms by x, we have

$$y=\frac{2-\dfrac{10}{x}}{1-\dfrac{4}{x}},$$

from which it is evident that as x increases without limit $10/x \rightarrow 0$, and $4/x \rightarrow 0$, and hence $y \rightarrow 2$.

Similarly, when $x \rightarrow -\infty$, that is, when x decreases without limit, $y \rightarrow 2$.

Therefore the graph approaches, both to the right and to the left, the line $y=2$ as an asymptote.

49. Value of a Fraction of the Form $\frac{\infty}{\infty}$. It is often necessary to consider the value of a fraction the terms of which are both infinite. Such a case was found in § 48, and most cases may be treated in the manner there shown.

For example, the fraction $(3x^2-2)/(2x^2-x+1)$ takes the form ∞/∞ when $x \rightarrow \infty$. But we may divide both terms of the fraction by the highest power of x in either term, and $(3x^2-2)/(2x^2-x+1)$ then becomes

$$\frac{3-\dfrac{2}{x^2}}{2-\dfrac{1}{x}+\dfrac{1}{x^2}}.$$

This fraction evidently approaches $\frac{3}{2}$ as a limit as x increases without limit.

50. Finding the Vertical and Horizontal Asymptotes. In order to find the vertical and horizontal asymptotes of a curve, we first solve the equation for y. If the result is a rational fraction in lowest terms, as in § 47, we see that

To every factor $x - a$ of the denominator there corresponds the vertical asymptote $x = a$ of the graph.

We also see from § 48 that

To every value a' which y approaches as a limit as x increases or decreases without limit there corresponds the horizontal asymptote $y = a'$ of the graph.

For example, in the equation

$$y = \frac{4x^2 - 12x - 16}{x^2 - 2x - 15}$$

the factors of the denominator are $x + 3$ and $x - 5$, and hence the lines $x = -3$ and $x = 5$ are vertical asymptotes; and as $x \longrightarrow \infty$ or as $x \longrightarrow -\infty$ we see that $y \longrightarrow 4$, from which it follows that the line $y = 4$ is a horizontal asymptote.

Certain other types of equation to which these methods do not apply, such as $y = \tan x$, $y = \log x$, and $y = a^x$, will be considered in Chapter XII on Higher Plane Curves.

51. Examination of an Equation. We are now prepared to summarize the steps to be taken in the examination of an equation under the following important heads:

1. Intercepts.
2. Symmetry with respect to points or lines.
3. Intervals of values of one variable for which the other is real or imaginary.
4. Asymptotes.

This examination is best illustrated by taking two typical equations, as in § 52.

Other methods of examining equations are given in the calculus.

52. Illustrative Examples. 1. Examine the equation $y=\frac{4x^2-12x-16}{x^2-2x-15}$, and draw the graph.

1. When $y=0$, then $4x^2-12x-16=0$, and $x=-1$ or 4, the two x intercepts. When $x=0$, then $y=1.1-$, the y intercept.

2. The graph is not symmetric with respect to OX, OY, or O.

3. As to intervals, y is real for all real values of x.

4. As to asymptotes, the vertical ones are $x=-3$ and $x=5$; the horizontal one is $y=4$, as was found in § 50.

We may now locate certain points suggested by the above discussion and draw the graph.

Convenient values for x are -8, -3.5, -2.5, 2, 4.5, 5.5, and 10. The two values of x near -3 and the two near 5 enable us to notice the graph's approach to the vertical asymptotes, while the values -8 and 10 suggest the approach to the horizontal asymptote.

In computing the values of y, notice that $y=\frac{4(x+1)(x-4)}{(x+3)(x-5)}$. Then, for example, when $x=-2.5$, we have

$$y=\frac{4(-\frac{3}{2})(-\frac{13}{2})}{\frac{1}{2}(-\frac{15}{2})}=-\frac{52}{5}=-10.4.$$

2. Examine the equation $y^2=\frac{4(x+1)(x-4)}{(x+3)(x-5)}$.

1. The second member being the same as in Ex. 1, the x intercepts are -1 and 4. The y intercepts are easily found to be ± 1.03.

2. The graph is symmetric with respect to OX.

3. As to intervals, when the sign of the fraction, which depends upon four factors, is negative, then y^2 is negative and hence y is imaginary. Let us follow x through the successive intervals determined by -3, -1, 4, 5, thus:

When $-\infty<x<-3$, all four factors are negative; hence y is real.

When $-3<x<-1$, the factors $x+1$, $x-4$, and $x-5$ are negative, but $x+3$ is positive; hence y is imaginary.

The student should continue the discussion of intervals.

4. The vertical asymptotes are $x=-3$, $x=5$; and since $y^2 \to 4$ when $x\to\infty$ or $-\infty$, the horizontal asymptotes are $y=2$, $y=-2$.

We may now locate certain points and draw the graph.

53. Limitations of the Method. The method of finding intervals and asymptotes described in this chapter is evidently limited to equations whose terms are positive integral powers of x and y, products of such powers, and constants. Moreover, the requirement that the equation be solved for one of the variables x and y in terms of the other is a simple one when the equation is of the first or second degree with respect to either of these variables, but it becomes difficult or impossible of fulfillment in the case of most equations of higher degrees.

Exercise 15. Graphs of Equations

Examine the following equations and draw the graphs:

1. $xy = 6.$
2. $xy = -6.$
3. $3xy = 6y + x.$
4. $3xy = 9y + 2x - 8.$
5. $2y + 16 = xy.$
6. $xy - 2y = x^2 - 16.$
7. $3x^2 - xy - 4x + y = 7.$
8. $\dfrac{xy^2}{4 - x} = 4.$
9. $y = \dfrac{9}{x^2 + 1} - \dfrac{1}{x^2 + 4}.$
10. $x^2y + y = 10.$
11. $(x^2 + y^2)(x + 4) = 12x^2.$
12. $y^2(5 + x) = -x^3.$
13. $y^2(3 + x) = x^2(3 - x).$
14. $y^2(x - 4) - x^2(x - 8) = 0.$
15. $x^2(y + 8) + y^3 = 0.$
16. $x^2y^2 = x^2 + y^2.$
17. $y^2 = \dfrac{9(x^2 - 2x - 8)}{x^2 - 6x + 5}.$
18. $y = \dfrac{x - 4}{2x^2 - 7x - 4}.$

19. A and B are two centers of magnetic attraction 10 units apart, and P is any point of the line AB. P is attracted by the center A with a force F_1 equal to $12/\overline{AP}^2$, and by the center B with a force F_2 equal to $18/\overline{BP}^2$. Letting $x = AP$, express in terms of x the sum s of the two forces, and draw a graph showing the variation of s for all values of x.

A ——10—— B — P

54. Degenerate Equation. It occasionally happens that when all the terms of an equation are transposed to the left member that member is factorable. In this case the equation is called a *degenerate equation.*

It is evident that the product of two or more factors can be 0 only if one or more of the factors are 0, and that any pair of values of x and y that makes one of the factors 0 makes the product 0 and satisfies the equation. It is therefore evident that all points (x, y) whose coordinates satisfy such an equation are precisely all the points whose coordinates make one or more factors 0. In other words,

The graph of a degenerate equation consists of the graphs of the several equations obtained by placing equal to 0 the several factors that contain either x or y, or both x and y.

For example, the graph of the equation $x^2 - xy - 3y + 3x = 0$, which may be written $(x - y)(x + 3) = 0$, consists of the two lines $x - y = 0$ and $x + 3 = 0$.

Exercise 16. Degenerate Equations

Draw the graphs of the following equations:

1. $(x - y)(y - 3) = 0$.
2. $x^2 = xy$.
3. $x^2 + xy = 0$.
4. $\dfrac{y(x^2 + y^2)}{36} = y$.
5. $y^3 + y^2 - 12y = 0$.
6. $(x - 4)^2 = (y + 2)^2$.
7. $(x - y)^2 = 9$.
8. $2x + 3y = \dfrac{2x^2 + 3xy}{2x + y}$.

9. If we divide both members of the cubic equation $(x - y)(x^2 - 6) = (x - y)x$ by $x - y$, we have the equation $x^2 - 6 = x$. How do the graphs of the two equations differ?

10. Show that the graph of the equation $xy = 0$ consists of the x axis and the y axis.

11. Draw the graph of the equation $(x^2 + y^2)^2 - 9x^2 = 9y^2$.

55. Intersections of Graphs. The complete solution of two simultaneous equations in x and y gives all pairs of values of the unknowns that satisfy both equations. Moreover, every pair of values of x and y that satisfies both equations locates a point of the graph of each equation. Hence we have the following rule:

To find the points of intersection of the graphs of two equations, consider the equations as simultaneous and solve for x and y.

If any one of the pairs of values of x and y has either x or y imaginary, there is no corresponding real point of intersection. In such a case, however, it is often convenient to speak of (x, y) as an *imaginary point of intersection.*

Exercise 17. Intersections of Graphs

Draw the graphs for each of the following pairs of equations and find their points of intersection:

1. $2x - y = 8$
   $x + y = 7$

2. $3x - 2y = 4$
   $2x + 5y = 9$

3. $x^2 + y^2 = 25$
   $x + y = 7$

4. $y^2 = 12x$
   $2y + x = 2$

5. $y^2 = 4x$
   $x^2 = \frac{1}{2}y$

6. $x^2 + y^2 = 16$
   $x^2 + y^2 - 4x = 12$

7. By finding the points of intersection, show that the line through (0, 5) and (3, 9) is tangent to the circle $x^2 + y^2 = 9$.

Consider the following pairs of equations with respect to the tangency of their graphs:

8. $4x + 5y = 41$
   $x^2 + y^2 = 41$

9. $5y + 3x = 14$
   $2xy + x - y = 8$

10. Find the points of intersection of the circles $x^2 + y^2 = 16$ and $x^2 + y^2 - 2x - 2y = 14$.

56. Variable and Constant. A quantity which is regarded as changing in value is called a *variable.* A quantity which is regarded as fixed in value is called a *constant.*

Frequently a symbol such as a, used to represent a constant, is free to represent any constant whatever, in which case it is called an *arbitrary constant.*

For example, the equation $(x-a)^2+(y-b)^2=r^2$ represents a circle (§ 36) with any center (a, b) and any radius r, and so we speak of a, b, and r as arbitrary constants and of x and y as variables.

57. Function. The student's work has already shown him that mathematics, pure and applied, is often concerned with two or more quantities which are mutually dependent upon each other according to some definite law.

For example, since $A=\pi r^2$, the area of a circle depends upon the radius; and, conversely, the radius may be said to depend upon the area.

Since the strength of a steel cable is given by the equation $S=kd^2$, where k is a constant, we see that the strength of a steel cable of given quality depends upon the diameter of the cable.

Since the number of beats per second of a pendulum l meters long is given approximately by the equation $n=1/l^2$, we see that the number of beats per second depends upon the length of the pendulum.

Since the distance through which a body falls from rest in t seconds is given by the equation $s=\frac{1}{2}gt^2$, where g is a constant, we see that the distance traversed depends upon the number of seconds of fall.

In each of these cases, when one of the variables is given, the other is easily determined.

If the first of two quantities depends upon the second in such a way as to be determined when the second is given, the first quantity is called a *function* of the second.

For example, x^2, $\sqrt{1-x}$, and $a^x \log x$ are functions of x. In the case of $A=\pi r^2$, A is a function of r, and r is a function of A.

Every equation in x and y defines a functional relation.

For example, the equation $4x^2-y^2=25$ defines y as a function of x, for $y=\sqrt{4x^2-25}$. It also defines x as a function of y.

58. Algebraic Function. A function obtained by applying to x one or more of the algebraic operations (addition, subtraction, multiplication, division, and the extraction of roots), a limited number of times, is called an *algebraic function of x.*

For example, x^3, $\frac{x}{x+1}$, and $3x^2 + \frac{1}{x} - \sqrt[3]{1-x}$ are algebraic functions of x.

59. Transcendent Function. If a function of x is not algebraic, it is called a *transcendent function of x.*

Thus, $\log x$ and $\sin x$ are transcendent functions of x.

60. Important Functions. Although mathematics includes the study of various functions, there are certain ones, such as $y = ax^n$, which are of special importance. For example, in § 57 we considered three equations, $A = \pi r^2$, $S = kd^2$, and $s = \frac{1}{2}gt^2$, all of which were of the same form, namely:

$$\text{one variable} = \text{constant} \times (\text{another variable})^2,$$

or

$$y = cx^2.$$

The most important functions which we shall study are algebraic functions of the first and second degrees.

61. Functional Relations without Equations. In making an experiment with a cooling thermometer the temperature t at the end of m minutes was found to be as follows:

$m =$	0	1	2	3	4	5	6	7	8	9
$t =$	98	65.6	44	29.5	19.4	13.2	8.9	5.9	4	2.1

We have no equation connecting t and m; yet t is a function of m, the values of t being found for given values of m by observation.

Such a functional relation may be represented by a graph in the usual manner.

62. Functional Notation. The symbol $f(x)$ is used to denote a function of x. It is read "f of x" and should not be taken to mean the product of f and x. It is often convenient to use other letters than f. Thus we may have $\phi(x)$, read "phi of x," $F(x)$, $R(x)$, and so on.

In a given function such as $f(x)$ we write $f(a)$ to mean the result of substituting a for x in the function.

For example, if $f(x) = x^2 - \dfrac{1}{x}$, then $f(1) = 1 - 1 = 0$, $f(3) = 8\frac{2}{3}$.

Exercise 18. Functions and Graphs

1. If $f(x) = (3 - x)^2$, find $f(1)$, $f(2)$, $f(3)$, and $f(4)$.

2. If $f(y) = 1/(1 - y)$, find $f(-1)$, $f(0)$, $f(\frac{1}{2})$, and $f(5)$.

3. Draw the graph showing the relation between t and m as given in the table in § 61, page 55.

4. From the following table draw two graphs, one showing the variation in per cents of population in cities of more than 2500 inhabitants, and the other that in rural communities:

Census	1880	1890	1900	1910	1920
Cities	29.5	36.1	40.5	46.3	52.3
Rural	70.5	63.9	59.5	53.7	47.7

5. From the following table draw a graph showing the variation of the time (t) of sunset corresponding to the day (d) of the year at a certain place, 4.49 meaning 4 hr. 49 min. P.M.:

$d =$	1	31	61	91	121	151	181	211	241	271	301	331	361
$t =$	4.49	5.20	5.53	6.23	6.52	7.18	7.29	7.13	6.36	5.48	5.05	4.40	4.45

Can points on the graph be found approximately for $d = 10$? for $d = 10\frac{1}{2}$? What kind of number must d be? Should the graph be a continuous curve? How many points are there on the entire graph?

Exercise 19. Review

Show that each of the following equations represents two straight lines and draw the graph of each equation:

1. $x^2 - y^2 = 0$.
2. $4x^2 - 9y^2 = 0$.
3. $2x^2 - 9y^2 = 0$.
4. $x^2 - 6xy + 8y^2 = 0$.
5. $x^2 + 5xy + 6y^2 = 0$.
6. $y^2 - xy - 12x^2 = 0$.
7. $3x^2 - 2xy - y^2 = 0$.
8. $6x^2 - 13xy + 6y^2 = 0$.

9. Two vertices of a triangle are $A(-4, 0)$ and $B(4, 0)$. The third vertex P moves so that $\overline{AP}^2 + \overline{BP}^2 = 64$. Find the equation of the locus of P and draw the locus.

10. The point P moves so that the slope of AP is half that of BP, where A is $(0, 0)$ and B is $(0, -6)$. Find the equation of the locus of P and draw the locus.

11. Find the equation of the locus of a point which moves so that its distance from the y axis is equal to the square of its distance from $(h, 0)$.

12. Find the equation of the locus of a point P which moves so that the sum of the squares of its distances from $(-3, 0)$, $(3, 0)$, and $(0, 6)$ is equal to 93. Draw the locus.

13. Find the equation of the locus of a point P which moves so that the angles APO and OPB are equal, where O is $(0, 0)$, A is $(-4, 0)$, and B is $(2, 0)$. Draw the locus.

If the angles are ϕ and ϕ', then $\tan\phi = \tan\phi'$.

14. Solve Ex. 13 when A is $(0, -4)$ and B is $(0, 2)$.

15. Given $A(-4, 0)$ and $B(4, 0)$, find the equation of the locus of a point P which moves so that the angle PAB is equal to twice the angle PBA. Draw the locus.

16. The equation of the straight line which bisects the angles between the axes in the first and third quadrants is $x = y$, and the equation of the line which bisects the angles between the axes in the other two quadrants is $x = -y$.

17. The fixed points A and B are on OX and OY respectively, and $OA = OB = 2a$. A point P moves so that the angles OPA and BPO are equal. Show that the equation of the locus of P is $(x-y)(x^2+y^2-2ax-2ay)=0$, and draw the locus.

18. The x intercept of the line $y = 6 - x$ is equal to the y intercept.

19. If the x intercept of the line $y = ax - 3$ is half the y intercept, find the value of a and draw the line.

20. If the graph of the equation $x^2 - xy + ay^2 = 23$ passes through the point $(3, -2)$, find the value of a.

21. If the graph of the equation $y = ax^2 + bx$ passes through the points $(-1, -3)$ and $(2, 18)$, find the values of a and b.

22. Show that the graph of the equation $y = \log_{10} x$ approaches, in the negative direction, the y axis as an asymptote, and draw the graph.

A table of logarithms (page 284) may be used for plotting points close together on this graph. But if the student will recall the well-known logarithms of the numbers 1, 10, 100, $\cdots$, and also of the numbers 0.1, 0.01, 0.001, $\cdots$, he will see the character of the graph clearly after plotting a few of the corresponding points.

23. Show that the graph of the equation $y = 2^x$ approaches the x axis as an asymptote, and draw the graph.

24. Draw the graph of the equation $x = \log_{10} y$ and compare it with the graph in Ex. 22.

25. Draw the graph of the equation $x = 2^y$ and compare it with the graph in Ex. 23.

26. Draw carefully in one figure the graphs of the equations $y = x$, $y = x^2$, $y = x^3$, and $y = x^4$, locating the points having x equal to 0, 0.3, 0.6, $\frac{3}{4}$, 1, 2, 3. Extend each graph by considering its symmetry.

A large unit of measurement, say 1 in., should be employed.

27. Consider Ex. 26 for the equations $y = x^{-1}$, $y = x^{-2}$, $y = x^{-3}$.

28. Draw the graphs of the equations $y^2 = x^4$ and $y^2 = x^3$

CHAPTER IV

THE STRAIGHT LINE

Problem. The Slope Equation

63. *To find the equation of the straight line when the slope and the y intercept are given.*

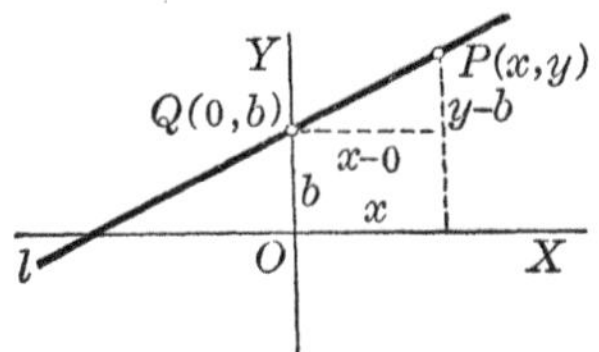

Solution. Let m be the given slope, b the given y intercept, and l the line determined by m and b.

Since b is the y intercept, l cuts OY in $Q(0, b)$.

Take $P(x, y)$ as any point on l. Then we have

$$\text{slope of } QP = \text{slope of } l;$$

that is, $$\frac{y-b}{x-0} = m; \qquad \text{§ 21}$$

whence $$\boldsymbol{y = mx + b.}$$

This equation is called the *slope equation* of the line.

When $m = 0$, then $y = b$, and the line is parallel to the x axis.

When $b = 0$, then $y = mx$, and the line passes through the origin.

When $m = b = 0$, then $y = 0$, and the line coincides with the x axis.

Since the y intercept is given, the line cuts the y axis, and the equation $y = mx + b$ cannot represent a line which is perpendicular to the x axis.

Theorem. Equation of any Straight Line

64. *The equation of any straight line of the plane is of the first degree in x and y.*

Proof. Since every line of the plane with the exception of lines parallel to the y axis has its slope m and its y intercept b, its equation is $y = mx + b$. § 63

Every line parallel to the y axis has for its equation $x = a$, where a is the distance from the y axis. § 35

Hence in every case the equation is of the first degree.

When we speak of a line of the plane we refer to the plane determined by the axes of x and y. We shall in general use the word *line* to mean a straight line where no ambiguity can result.

Exercise 20. The Slope Equation

Given the following conditions relating to a straight line, draw the line and find its equation:

1. The slope is 5 and the y intercept is 7.
2. Each of the two intercepts is 10.
3. The slope is -4 and the point $(0, 6)$ is on the line.
4. The line contains $(6, -3)$ and the y intercept is 8.
5. The line passes through the points $(4, 2)$ and $(0, -2)$.
6. The x intercept is -5 and the y intercept is 3.
7. The line passes through the points $(2, 0)$ and $(2, 6)$.

Write each of the following equations in the slope form, draw each line, and find the slope and the two intercepts:

8. $3y = 6x + 10$.
9. $x + y = 6$.
10. $3y - 2x - 12 = 0$.
11. $12 - 2y = 9x$.

12. The lines $3x + 4y = 24$ and $8y + 6x = 45$ are parallel.

13. The lines $2x - 5y = 20$ and $5x + 2y = -8$ intersect at right angles on the y axis.

THEOREM. EQUATION OF THE FIRST DEGREE

65. *Every equation of the first degree in x and y represents a straight line.*

Proof. Taking A, B, and C as arbitrary constants,

$$Ax + By + C = 0$$

represents any equation of the first degree in x and y.

It is evident that A, B, or C, or both A and C, or both B and C, may be 0; but both A and B cannot be 0, for C would then be 0 and we should have only the identity $0 = 0$.

If $B = 0$, the equation becomes $Ax + C = 0$, or $x = -C/A$, the equation of a line parallel to the y axis.

If $B \neq 0$, we may divide by B and write the equation in the form

$$y = -\frac{A}{B}x - \frac{C}{B},$$

which is the equation of a line with slope $m = -A/B$ and y intercept $b = -C/B$. § 63

Therefore in every case the equation $Ax + By + C = 0$ represents a straight line.

66. COROLLARY 1. *The slope of the line is $-A/B$.*

67. COROLLARY 2. *The line passes through the origin when and only when the equation has no constant term.*

For $Ax + By = 0$ is satisfied by the coordinates of the origin $(0, 0)$; but $Ax + By + C = 0$ is not satisfied by $x = 0$, $y = 0$ unless $C = 0$.

68. Determining a Line. Although a line is determined by its slope and its y intercept, it is also determined by other conditions, as by two points, by its slope and one point, and so on. A few of the resulting equations of a line are used so frequently as to be regarded as fundamental. The slope equation $y = mx + b$ is a fundamental equation, and we shall now consider others.

PROBLEM. THE POINT-SLOPE EQUATION

69. *To find the equation of the straight line through a given point and having a given slope.*

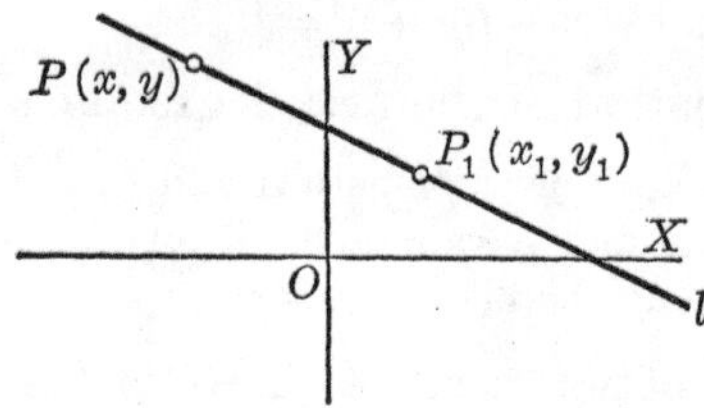

Solution. Let $P_1(x_1, y_1)$ be the given point, m the given slope, and l the line determined by them.

Take $P(x, y)$ as any point on l. Then we have

$$\text{slope of } PP_1 = \text{slope of } l;$$

that is,
$$\frac{y - y_1}{x - x_1} = m; \qquad \S\ 21$$

whence
$$y - y_1 = m(x - x_1).$$

This equation is called the *point-slope equation* of the line.

Exercise 21. The Slope and Point-Slope Equations

Find the intercepts and slope of each of the following lines, and by the aid of the intercepts draw each line:

1. $6x + 5y = 30.$

2. $4x - 18 = 3y.$

3. $3x = 4y - 12.$

4. $2y + 7x = -9.$

5. $x - 3y - 8 = 0.$

6. $2y - 4 + 5x = 0.$

Draw each of the following lines and find the slope:

7. $y = 8x.$

8. $x = 2y.$

9. $3x = 7y.$

10. $2y = 5x.$

11. $4x + 3y = 0.$

12. $x + \frac{3}{2}y = 0.$

Find the equations of the lines subject to the following conditions, and draw each line:

13. Through $(4, -2)$ and having the slope -3.

14. Through $(-1, 2)$ and parallel to the line $3x - y = 7$.

15. Through $(4, 1)$ and parallel to the line through the points $(5, 2)$ and $(4, 4)$.

16. Having the slope 2 and the x intercept -5.

17. Having the slope m and the x intercept a.

18. Through $(3, -2)$ and having the intercepts equal.

Determine the value of k, given that:

19. The line $2x - ky = 9$ passes through $(-3, 1)$.

20. The slope of the line $2kx - 7y + 4 = 0$ is $\frac{1}{2}$.

21. The line $3x + ky = 3$ has equal intercepts.

22. The line $y - 2 = k(x - 4)$ has equal intercepts.

23. The line through the point $(4, -1)$ with slope k has the y intercept 10.

24. Draw the line $2x - ky = 9$ for the values $k = 1, 2, \frac{1}{3}, 0, -1, -3$, and -6.

25. Find the equation of the line through $(4, -2)$ which makes with the axes a triangle of area 2 square units.

26. Temperatures on the Fahrenheit and centigrade scales are connected by the linear relation $F = aC + b$. Knowing that water freezes at 32° F. or 0° C., and boils at 212° F. or 100° C., find the values of a and b and draw the graph.

Should this graph extend indefinitely in both directions?

27. A certain spiral spring is stretched to the lengths 3.6 in. and 4.4 in. by weights of 8 lb. and 12 lb. respectively. Assuming that varying weights w and the corresponding lengths l are connected by a linear relation, find that relation and draw the graph. What is the length of the unstretched spring?

Should this graph extend indefinitely in both directions?

Problem. The Two-Point Equation

70. *To find the equation of the straight line through two given points.*

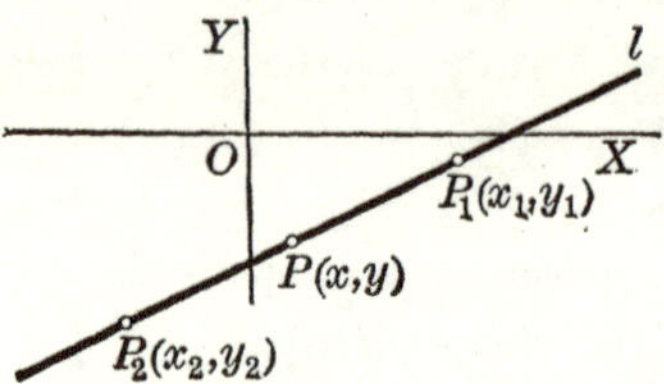

Solution. Let $P_1(x_1, y_1)$ and $P_2(x_2, y_2)$ be the given points, l the given line, and $P(x, y)$ any point on l. Then the equation of l is found from the usual condition that

$$\text{slope of } PP_1 = \text{slope of } P_1P_2;$$

that is,

$$\frac{y - y_1}{x - x_1} = \frac{y_2 - y_1}{x_2 - x_1}. \qquad \S\, 21$$

This equation is called the *two-point equation* of the line and is often written $y - y_1 = \frac{y_2 - y_1}{x_2 - x_1}(x - x_1)$.

71. Determinant Form of the Two-Point Equation. Students who are familiar with the determinant notation should notice that the two-point equation may conveniently be written in the following form:

$$\begin{vmatrix} x & y & 1 \\ x_1 & y_1 & 1 \\ x_2 & y_2 & 1 \end{vmatrix} = 0.$$

This equation, when the determinant is expanded, is identical with the equation of § 70 when cleared of fractions, as the student may easily verify.

Without expanding the determinant the student may observe the effect of substituting x_1 for x, and y_1 for y.

72. Parallel Lines. Since two lines are parallel if they have the same slope, the equations $y = mx + b$ and $y = m'x + b'$ represent parallel lines when $m = m'$.

The equations $Ax + By + C = 0$ and $A'x + B'y + C' = 0$ represent parallel lines if

$$-\frac{A}{B} = -\frac{A'}{B'}; \text{ that is, if } \frac{A}{B} = \frac{A'}{B'}, \text{ or if } \frac{A}{A'} = \frac{B}{B'}. \quad \S\,66$$

For example, the lines $3x - 2y = 7$ and $6x - 4y - 39 = 0$ are parallel, since $\frac{3}{6} = \frac{-2}{-4}$.

73. Perpendicular Lines. Since two lines with slopes m and m' are perpendicular to each other when $m' = -1/m$ (§ 24), two such lines are represented by the equations $y = mx + b$ and $y = -\frac{1}{m}x + b'$.

The equations $Ax + By + C = 0$ and $A'x + B'y + C' = 0$ represent two lines perpendicular to each other if

$$-\frac{A}{B} = -\frac{1}{-A'/B'}; \text{ that is, if } \frac{A}{B} = -\frac{B'}{A'}.$$

This condition is often written $AA' + BB' = 0$.

If $Ax + By + C = 0$ is the equation of a certain line, then the line $Ax + By + K = 0$ is parallel to the given line and the line $Bx - Ay + H = 0$ is perpendicular to the given line.

74. Lines at any Angle. The angle θ from one line, $y = mx + b$, to another, $y = m'x + b'$, may be found from the relation

$$\tan\theta = \frac{m' - m}{1 + mm'}. \quad \S\,25$$

In the case of the pair of lines $Ax + By + C = 0$ and $A'x + B'y + C' = 0$ we have $m = -A/B$ and $m' = -A'/B'$;

whence
$$\tan\theta = \frac{A'B - AB'}{AA' + BB'}.$$

Exercise 22. Two-Point Equation, Parallels, Perpendiculars

1. Find the equation of the line through $(4, -3)$ and parallel to the line $5x + 3y = 8$.

The slope of the given line is $-\frac{5}{3}$, and hence § 69 applies.

2. Find the equation of the line through $(4, -3)$ and perpendicular to the line $6x - 4y = 13$.

The slope of the given line is evidently $\frac{3}{2}$, and hence the slope of the required line is $-\frac{2}{3}$. Then apply § 69.

Find the equations of the lines through the following points:

3. $(4, 6)$ and $(3, -1)$.

4. $(5, 2)$ and $(-4, -3)$.

5. $(7, -2)$ and $(0, 0)$.

6. $(-6, 1)$ and $(2, 0)$.

7. $(-1, 3)$ and the intersection of $x - y = 3$ with $x + 2y = 9$.

8. $(2, 4)$ and the intersection of $2x + y = 8$ with the x axis.

9. The common points of the curves $y^2 = x$ and $x^2 = y$.

10. Find the equation of the line through the point $(4, -2)$ and parallel to $4x - y = 2$.

11. Find the equation of the line through the point $(2, 2)$ and parallel to the line through $(-1, 2)$ and $(3, -2)$.

12. Find the equation of the line through the point $(-3, 0)$ and perpendicular to $7x - 3y = 1$.

13. Find the equation of the line through the point $(-1, -3)$ and perpendicular to the line of Ex. 4.

Draw the triangle whose vertices are $A(5, 3)$, $B(7, -1)$, $C(-1, 5)$, *and find the equations of the following lines:*

14. The three sides of the triangle.

15. The three medians of the triangle.

16. The three altitudes of the triangle.

17. The three perpendicular bisectors of the sides.

18. Given that the line $Ax + By + 10 = 0$ is parallel to the line $3x + y = 7$ and meets the line $x + y = 7$ on the x axis, find the values of A and B.

Through each of the vertices $A(5, 3)$, $B(7, -1)$, $C(-1, 5)$ of the triangle ABC draw a line parallel to the opposite side, and for the triangle thus formed consider the following:

19. The equation of each of the three sides.

20. The coordinates of each of the vertices.

21. The mid points of the sides are A, B, and C.

22. The three medians are concurrent.

Show that the intersection of any two medians is on the third.

23. The three altitudes are concurrent.

24. The perpendicular bisectors of the sides are concurrent.

25. Draw the line from the origin to $A(7, -3)$ and find the equation of the line perpendicular to OA at A.

26. Draw the circle $x^2 + y^2 = 34$, show that it passes through $A(5, 3)$, and find the equation of the tangent at A.

27. Draw the circle $x^2 + y^2 - 2x + 4y - 5 = 0$ and find the equation of the tangent at $(2, 1)$.

28. Given that the line $kx - y = 4$ is perpendicular to the line $kx + 9y = 11$, find the value of k.

29. Locate $A(12, 5)$, on OA lay off OB equal to 10, and find the equation of the line perpendicular to OA at B.

30. Find the distance from the line $12x + 5y - 26 = 0$ to the origin.

31. Given that three vertices of a rectangle are $(2, -1)$, $(7, 11)$, and $(-5, 16)$, find the equations of the diagonals.

32. Find the point Q such that $P(3, 5)$ and Q are symmetric with respect to the line $y + 2x = 6$.

Denote Q by (a, b), and find a and b from the two conditions that PQ is perpendicular to the line and the mid point of PQ is on the line.

Problem. The Normal Equation

75. *To find the equation of the straight line when the length of the perpendicular from the origin to the line and the angle from the x axis to this perpendicular are given.*

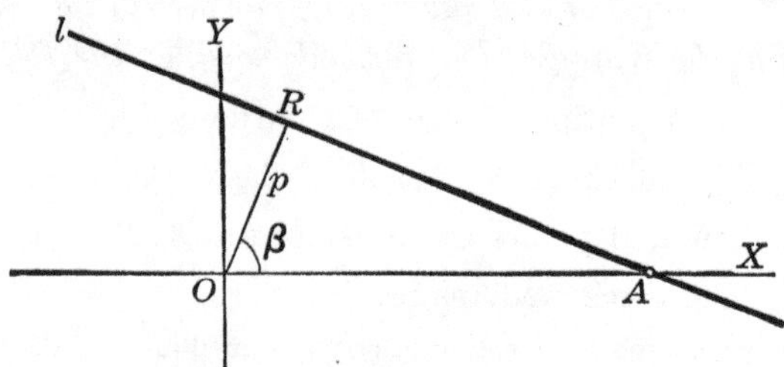

Solution. Let l be the given line and OR the perpendicular from O to l. Denote the angle XOR by β and the length of OR by p.

Since the slope of OR is $\tan\beta$, then the slope of l is $-1/\tan\beta$, or $-\cos\beta/\sin\beta$. § 73

And since in the triangle OAR, $\cos\beta = p/OA$, it follows that $OA = p/\cos\beta$, and hence A is the point $(p/\cos\beta,\ 0)$.

Therefore the required equation of l is

$$y - 0 = -\frac{\cos\beta}{\sin\beta}\left(x - \frac{p}{\cos\beta}\right), \qquad \text{§ 69}$$

or

$$\boldsymbol{x\cos\beta + y\sin\beta - p = 0.}$$

This equation is called the *normal equation* of the line.

Since p is positive, *the constant term $-p$ in the normal equation is always negative.* The angle β may have any value from 0° to 360°.

76. Line through the Origin. If l passes through O, $p = 0$ and the equation of l is $x\cos\beta + y\sin\beta = 0$.

In this equation we shall always take $\beta < 180°$, so that $\sin\beta$, the coefficient of y, is always positive.

PROBLEM. THE INTERCEPT EQUATION

77. *To find the equation of the line in terms of the intercepts.*

Solution. Denoting the x intercept of the given line l by a and the y intercept by b, it is evident that l cuts the x axis at $A(a, 0)$ and the y axis at $B(0, b)$. The student may now show that the equation of l is

$$\frac{x}{a}+\frac{y}{b}=1.$$

This equation is called the *intercept equation* of the line.

78. Changing the Form of the Equation. The equation of the straight line has now appeared in six different forms:

1. General	$Ax+By+C=0$	§ 65
2. Slope	$y=mx+b$	§ 63
3. Point-slope	$y-y_1=m(x-x_1)$	§ 69
4. Two-point	$y-y_1=\frac{y_2-y_1}{x_2-x_1}(x-x_1)$	§ 70
5. Normal	$x\cos\beta+y\sin\beta-p=0$	§ 75
6. Intercept	$\frac{x}{a}+\frac{y}{b}=1$	§ 77

Each of these forms may be reduced to the general form, and the general form may be written in each of the other forms, as is illustrated in the following examples:

1. Change the slope form to the general form.

Transposing, we have the general form $-mx+y-b=0$.

2. Change the general form to the slope form.

Dividing by B and transposing, we have the result.

3. Change the general form to the intercept form.

Transpose C, divide each term by $-C$, and divide both terms of each fraction by the coefficient of the numerator.

79. Changing the General Form to the Normal Form. The change from the general form to the normal form is not so simple as the changes in the three examples of § 78.

In the normal form, $x \cos \beta + y \sin \beta - p = 0$, it is evident that neither $\cos \beta$ nor $\sin \beta$ can exceed 1, and hence that each is, in general, a proper fraction. Therefore, in order to change $Ax + By + C = 0$ to the normal form we divide by some number k, the result being

$$\frac{A}{k}x + \frac{B}{k}y + \frac{C}{k} = 0.$$

In order to determine k we first note that A/k is to be $\cos \beta$, and B/k is to be $\sin \beta$. Then since $\cos^2 \beta + \sin^2 \beta = 1$, k must be so chosen that

$$\left(\frac{A}{k}\right)^2 + \left(\frac{B}{k}\right)^2 = 1.$$

Therefore $$\boldsymbol{k = \pm\sqrt{A^2 + B^2}},$$

and the sign of k must be such that C/k is negative (§ 75) or, when $C = 0$, so that B/k is positive (§ 76).

Hence, *to change $Ax + By + C = 0$ to the normal form divide by $\sqrt{A^2 + B^2}$ preceded by the sign opposite to that of C, or, when $C = 0$, by the same sign as that of B.*

The result is

$$\frac{A}{\pm\sqrt{A^2 + B^2}}x + \frac{B}{\pm\sqrt{A^2 + B^2}}y + \frac{C}{\pm\sqrt{A^2 + B^2}} = 0,$$

from which it appears that

$$\cos \beta = \frac{A}{\pm\sqrt{A^2 + B^2}}, \quad \sin \beta = \frac{B}{\pm\sqrt{A^2 + B^2}}, \quad p = \frac{C}{\pm\sqrt{A^2 + B^2}}.$$

For example, to change $4x + 3y + 15 = 0$ to the normal form we divide by $-\sqrt{4^2 + 3^2}$, or -5. We then have $-\frac{4}{5}x - \frac{3}{5}y - 3 = 0$, in which $\cos \beta = -\frac{4}{5}$, $\sin \beta = -\frac{3}{5}$, and $p = 3$. It is therefore evident that β is an angle in the third quadrant.

Exercise 23. The Normal Equation

1. Find the distance from (0, 0) to the line $4x - 4y = 9$.

Since only the length of p is required, we may divide at once by $\sqrt{16+16}$, or $4\sqrt{2}$, without considering the sign.

The normal equation is used chiefly in problems involving the distance of a line from the origin.

2. Find the equation of the line which is 7 units from the origin and has the slope 3.

Since $m = 3$, the equation is $y = 3x + b$, which becomes, in normal form, $(-3x + y - b)/(\pm\sqrt{10}) = 0$, from which $p = b/(\pm\sqrt{10}) = 7$; whence $b = \pm 7\sqrt{10}$. Hence two lines satisfy the given condition; namely, $y = 3x + 7\sqrt{10}$ and $y = 3x - 7\sqrt{10}$.

Write the following equations in slope, intercept, and normal forms and find the values of m, b, a, p, and β:

3. $6x - 8y = 35$.

4. $2x = 3y - 6$.

5. $2y - 5x = 20$.

6. $y = 6(x - 3)$.

7. $3y - 6x + 10 = 0$.

8. $2y - 3x = 0$.

Find the distance from (0, 0) to each of the following lines:

9. $5y = 12x - 91$.

10. $y = 7x - 3$.

11. $\frac{x}{2} + \frac{y}{5} = 1$.

12. $y = mx + b$.

13. $y - 2 = m(x - 5)$.

14. $\frac{y-k}{x-l} = m$.

Determine k so that the distances from the origin to the following lines shall be as stated:

15. $y = kx + 9$; $p = 6$.

16. $y + 1 = k(x - 3)$; $p = 2$.

Find the equations of the lines subject to these conditions:

17. Slope -3; 10 units from the origin.

18. Through (2, 5); 1 unit from the origin.

19. Midway between O and the line $3x - 4y - 30 = 0$.

80. Two Essential Constants. Although the equation $Ax + By + C = 0$, which represents any line of the plane, has three arbitrary constant coefficients, we may divide by any one of these coefficients and reduce the number to two. For example, if we divide by C, the equation becomes

$$\frac{A}{C}x + \frac{B}{C}y + 1 = 0,$$

from which it appears that only two arbitrary constants, the coefficients of x and y in this form of the equation, are essential to determine the equation.

We therefore say that *the equation of the straight line involves two essential constants.*

81. Finding the Equation of a Line. To find the equation of the line determined by two given conditions we choose one of the six fundamental forms (§ 78) to represent the line. The problem is to express the two given conditions in terms of the two required constants in the chosen form and thus to find these constants.

For example, find the equation of a line, given that its y intercept is twice its x intercept and that its distance from the origin is 12 units.

Suppose that the form $y = mx + b$ is chosen. Then by § 45 the y intercept is b and the x intercept is $-b/m$. Therefore the first condition is that $b = -2b/m$.

Since by § 79 the distance from O to $y = mx + b$ is $b/(\pm\sqrt{1 + m^2})$, the second condition is that $b/(\pm\sqrt{1 + m^2}) = 12$.

Solving the equations that represent these conditions, we have

$$m = -2,$$

and

$$b = \pm 12\sqrt{5}.$$

The required equation is, therefore,

$$y = -2x \pm 12\sqrt{5}.$$

That is, there are two lines, as is evident from the conditions.

82. System of Straight Lines. All lines which satisfy a single condition are said to form a *system of lines* or a *family of lines.*

Thus, $y = 6x + b$ represents the system of lines with slope 6. These lines are evidently parallel, and any particular one is determined when the arbitrary constant b is known.

Similarly, $y - 6 = m(x - 3)$ represents the system of lines through (3, 6). In this case we have what is known as a *flat pencil of lines,* and any one of them is determined when m is known.

Any equation of the line which involves only a single arbitrary constant, such as $kx - 2y + 5 = 0$, represents a system of lines, each line of the system corresponding to some particular value of k. If a second condition is given, it is evident that k is thereby determined.

Exercise 24. Equations of Lines

Given $y = mx + b$, the equation of the line l, determine m and b under the following conditions:

1. Line l is 4 units from O, and the x intercept is -8.
2. Line l has equal intercepts and passes through $(5, -3)$.
3. Line l passes through O, and the angle from l to the line $4x - 7y - 28 = 0$ is 45°.
4. Line l passes through $A(10, 2)$ and cuts the x axis in C, AC being equal to 18.
5. Line l passes through $A(4, 6)$ and cuts the axes in B and C in such way that A is the mid point of BC.

Given $\dfrac{x}{a} + \dfrac{y}{b} = 1$, the equation of the line l, determine a and b under the following conditions:

6. Line l is 4 units from O, and the x intercept is -8.
7. Line l has equal intercepts and passes through $(5, -3)$.

Given $x \cos \beta + y \sin \beta - p = 0$, the equation of the line l, determine the constants under the following conditions:

8. Line l passes through $(-4, 12)$, and $\beta = 45°$.

9. Line l passes through $(14, -2)$, and $p = 10$.

10. Line l touches the circle $x^2 + y^2 = 16$, and $\tan \beta = \frac{4}{3}$.

Given $y - 4 = m(x + 6)$, the equation of a line l passing through $(-6, 4)$, determine m under the following conditions:

11. Line l is 2 units from the origin.

12. The intercepts of the line l are numerically equal, but have unlike signs.

13. The x intercept of the line l is -9.

Find the equations of the following lines:

14. The line passing through O and meeting the line $x+y=7$ at the point P such that $OP = 5$.

15. The line passing through the point $(1, 7)$ and tangent to the circle $x^2 + y^2 = 25$.

16. The line cutting the lines $x + 2y = 10$ and $2x - y = 10$ in the points A and B such that the origin bisects AB.

17. The line having the slope 6 and cutting the axes at A and B so that $AB = \sqrt{37}$.

18. Find the distance from the line $4x + 3y = 5$ to the point $(6, 9)$.

First find the equation of the line through the point $(6, 9)$ parallel to the given line, and then find the distance between these lines.

19. Find the distance from the line $5x + 12y = 60$ to the point $P(8, 6)$ by the following method: Draw the perpendicular PK to the line; draw the ordinate MP of P, and denote by Q its intersection with the line; then find KP from the right triangle PKQ, in which the angle PKQ is equal to the angle β for the line $5x + 12y = 60$.

Problem. Distance from a Line to a Point

83. *Given the equation of a line and the coordinates of a point, to find the distance from the line to the point.*

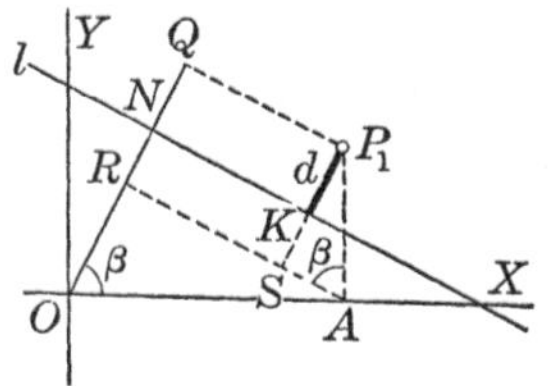

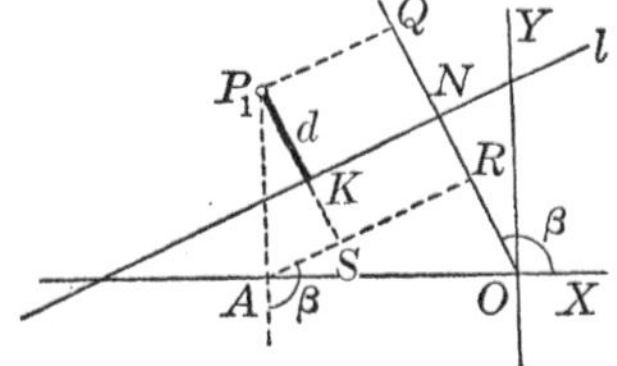

Solution. Let l be the given line and let its equation in the normal form be $x \cos \beta + y \sin \beta - p = 0$, in which β and p are known; and let $P_1(x_1, y_1)$ be the given point. We are to compute d, the distance KP_1, from l to P_1. Draw ON perpendicular to l. Then $ON = p$. Draw P_1A perpendicular to OX, and P_1Q and AR perpendicular to ON.

Then
$$\begin{aligned} d &= NQ = OQ - ON \\ &= OQ - p \\ &= OR + RQ - p. \end{aligned}$$

To find OR and RQ, and thus to find d, we first see that
$$\begin{aligned} OR &= x_1 \sin RAO \\ &= x_1 \cos \beta. \end{aligned}$$

Producing P_1K to meet AR at S, we have $SP_1 = RQ$, $AP_1 = y_1$, and both AR and AP_1 perpendicular to the sides of β. Therefore, whatever the shape of the figure, one of the angles formed by AR and AP_1 is equal to β, and hence
$$RQ = y_1 \sin \beta.$$

Therefore
$$OR + RQ = x_1 \cos \beta + y_1 \sin \beta,$$
and
$$\boldsymbol{d = x_1 \cos \beta + y_1 \sin \beta - p.}$$

84. Distance from a Line to a Point. The result found in the preceding problem (§ 83) may now be stated in the following rule:

To find the distance from a given line to a given point $P_1(x_1, y_1)$, *write the equation of the line in the normal form and substitute* x_1 *for* x *and* y_1 *for* y. *The left-hand member of the resulting equation expresses the distance required.*

If, as is commonly the case, the equation is given in the general form $Ax + By + C = 0$, we have

$$\boldsymbol{d = \frac{Ax_1 + By_1 + C}{\pm\sqrt{A^2 + B^2}}},$$

where the sign of $\sqrt{A^2 + B^2}$ is chosen by the rule of § 79.

Thus, the distance from the line $4x - 5y + 10 = 0$ to $P(-6, 3)$ is

$$d = \frac{4(-6) - 5 \cdot 3 + 10}{-\sqrt{4^2 + (-5)^2}} = \frac{29}{\sqrt{41}};$$

while the distance from the same line to $P(5, 1)$ is

$$d = \frac{4 \cdot 5 - 5 \cdot 1 + 10}{-\sqrt{4^2 + (-5)^2}} = -\frac{25}{\sqrt{41}}.$$

One of these distances is positive, while the other is negative.

85. Sign of *d*. Only the length of d, without regard to the sign, is usually required, although the sign is occasionally essential. In the work on page 75, d denotes the *directed length* KP_1; that is, the distance measured from l towards P_1. In other words, $d = OQ - ON$, where ON is always positive (§ 75). Hence d is positive when $OQ > ON$, and negative when $OQ < ON$.

That is, *d is positive when it has the same direction as the perpendicular from O to l, and is negative when it has the opposite direction.*

If l passes through O, the direction of p is from O along the upper part of the perpendicular, since $\beta < 180°$.

86. Distance to a Point from a Line Parallel to an Axis. To find the distance to a given point $P_1(x_1, y_1)$ from a line parallel to an axis, say from the line $x = a$, is a special case of § 83. In this case the distance is easily seen to be

$$d = OQ - a = x_1 - a.$$

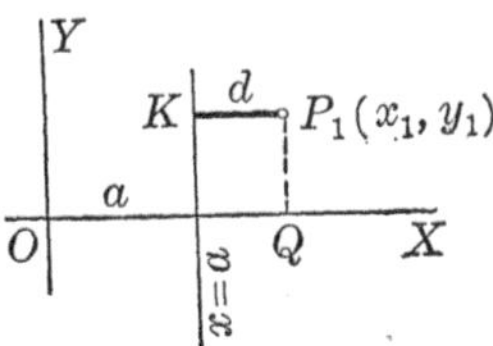

Similarly, the distance from a line parallel to the x axis, say the line $y = b$, to the point (x_1, y_1) is $d = y_1 - b$.

Exercise 25. Distance from a Line to a Point

Find the distances from the following lines to the points specified in each case:

1. Line $8x + 6y = 55$; points $(10, 2)$, $(9, 6)$, $(4, 5)$, $(4, -1)$, $(-1, 1)$.
2. Line $5x - 12y = 26$; points $(13, 0)$, $(-2, -3)$, $(0, 0)$.
3. Line $y = \frac{4}{3}x - 5$; points $(3, 6)$, $(6, 1)$, $(2, 2)$, $(-1, 3)$, $(-2, -6)$.
4. Line $y - 2x + 7 = 0$; points $(-1, 1)$, $(0, -1)$, $(6, -3)$, $(-1, 3)$.
5. Line $y = 5x$; points $(4, 1)$, $(2, -2)$, $(-3, 0)$, $(-1, -5)$.
6. Line $y = mx + 4$; points $(0, 0)$, $(1, 1)$, $(-4, 2)$, (a, b).
7. Line $y = mx + b$; points $(4, -1)$, $(3, 2)$, $(a, 1)$, (k, l).
8. Line $y - 6 = m(x + 1)$; points $(\sqrt{1 + m^2}, 6 + \sqrt{1 + m^2})$, $(1, 4)$, $(-2, 1)$.

Find the altitudes of the triangles whose vertices are:

9. $(4, 1)$, $(8, -2)$, $(1, -3)$.
10. $(-5, -1)$, $(5, \frac{5}{4})$, $(-8, 11)$.

Find the altitudes of the triangles whose sides are:

11. $8y + x + 34 = 0$, $x - y = 2$, $2x + y = 7$.
12. $3x = 4y$, $6x + 8y - 5 = 0$, $5x - 12y - 17 = 0$.

Draw the line represented by each of the following equations, locate C, draw the circle with center C and tangent to the line, and find the radius and the equation of the circle:

13. $6x - 8y - 25 = 0$; $C(4, 3)$. **15.** $2y = 5x$; $C(-1, 4)$.

14. $y = 3x + 10$; $C(3, -2)$. **16.** $y = x$; $C(0, 6)$.

Find the value of m in each of the following cases:

17. The line $y = 5mx + 8$ is 5 units from the point (4, 5).

18. The line $y = mx + 10$ is tangent to the circle with radius 6 and center (0, 0).

19. The line $y - 1 = m(x - 7)$ is tangent to the circle $x^2 + y^2 = 25$.

20. The lines $8x + 6y = 11$ and $8mx - 8y = 7$ are equidistant from the point $(1, -2)$.

Find the equations of the lines described as follows:

21. Perpendicular to the line $7x + y = 5$ and 4 units from (6, 1).

22. Passing through $(-1, -4)$ and 6 units from (3, 2).

23. Parallel to the line $5x - 12y = 17$ and tangent to the circle $x^2 + y^2 - 8x + 12y = 12$.

24. Find the values of m and b if it is given that the line $y = mx + b$ is 6 units from the point (4, 1) and 2 units from the origin.

To find m and b from the resulting equations first divide the members of one equation by those of the other.

25. Find the values of m and b if the line $y = mx + b$ is equidistant from the points (4, 1), (8, 2) and $(5, -3)$.

26. Find the values of m and b if the line $y = mx + b$ passes through the point $(12, -6)$ and the x intercept is $\frac{5}{3}$ of the distance from the origin.

27. Find the values of a and b if the line $x/a + y/b = 1$ passes through (3, 4) and is 3 units from (6, 5).

28. Find the values of a and b if the line $x/a + y/b = 1$ has equal intercepts and is $4\sqrt{2}$ units from $(5, -2)$.

29. Find the values of a and b if the line $x/a + y/b = 1$ is 2 units from $(a, -3)$ and the x intercept exceeds the y intercept by 5.

30. Find the equation of the line which is 4 units from the point $(-2, 6)$, the angle from the line to the line $7x + y = 0$ being 45°.

31. Find the equation of the line which is as far from $(4, 1)$ as from $(1, 4)$ and is parallel to the line $5x - 2y = 1$.

32. Find the equation of the line which is equidistant from $(2, -2)$, $(6, 1)$, and $(-3, 4)$.

33. Find the equation of the line the perpendicular to which from $(1, -6)$ is bisected by $(4, -2)$.

Given the following conditions, find k and l:

34. Given that (k, l) is equidistant from the lines $4x = 3y - 12$, $9x + 12y = 18$, and $6x - 8y = 25$.

In the formula of § 84 the signs $\pm$ before the radical correspond to distances on opposite sides of the line. Here we have $d = \pm d' = \pm d''$, giving four cases. The student should complete one of them.

35. Given that (k, l) is -4 units from the line $12x - 5y = 49$ and -3 units from the line $4y = 3x + 24$.

36. Given that (k, l) is equidistant from the lines $3x - y = 0$ and $x + 3y = 4$, and that $k = l$.

37. Given that (k, l) is equidistant from the lines $4x - 2y = 15$ and $x - 2y = 5$, and also equidistant from the lines $x + y = 6$ and $x = y$.

38. Given that (k, l) is on the line $2y + y + 2 = 0$, and is 4 units from the line $3x - 4y = 10$.

39. Find the equation of the locus of a point which moves so that its distance from the line $6x + 2y = 13$ is always twice its distance from the line $y = 3x + 8$.

Theorem. Lines through the Intersection of Lines

87. *Given* $A_1x + B_1y + C_1 = 0$ *and* $A_2x + B_2y + C_2 = 0$, *the equations of the lines* l_1 *and* l_2 *respectively, then the equation* $A_1x + B_1y + C_1 + k(A_2x + B_2y + C_2) = 0$, *where* k *is an arbitrary constant, represents the system of lines* l_3 *through the intersection of the lines* l_1 *and* l_2.

Proof. The equation of l_3 is of the first degree in x and y, whatever may be the value of k, and hence represents a straight line. § 65

Furthermore, l_3 passes through the intersection of l_1 and l_2, since the pair of values which satisfies the equations of l_1 and l_2 also satisfies the equation of l_3.

Moreover, k can be determined by the condition that l_3 passes through any other point (x_1, y_1) of the plane, for the substitution of x_1 for x and of y_1 for y in the equation of l_3 leaves k the only unknown quantity.

Hence the equation $A_1x + B_1y + C_1 + k(A_2x + B_2y + C_2) = 0$ represents, for various values of k, all the lines passing through the intersection of l_1 and l_2.

88. Illustrative Examples. 1. Find the equation of the line through the intersection of the lines $3x - 2y = 4$ and $y = 4x - 7$, and through $(4, -2)$.

By § 87 the equation is $3x - 2y - 4 + k(y - 4x + 7) = 0$. Since the line passes through $(4, -2)$, by substituting 4 for x and -2 for y we find that $k = \frac{12}{11}$. Substituting $\frac{12}{11}$ for k and simplifying we have $3x + 2y = 8$ as the required equation.

2. Find the equation of the line through the intersection of the lines $4x + y - 11 = 0$ and $3x - y = 3$, and perpendicular to the line $x + 10y = 7$.

In $4x + y - 11 + k(3x - y - 3) = 0$, k is to be so chosen that the slope is 10; that is, $-(4 + 3k)/(1 - k) = 10$; whence $k = 2$. Substituting 2 for k and simplifying, we have $10x - y - 17 = 0$.

THEOREM. EQUIVALENT EQUATIONS

89. *The equations* $Ax + By + C = 0$ *and* $A'x + B'y + C' = 0$ *represent the same line if and only if the corresponding coefficients are proportional.*

Proof. The equations represent the same line if and only if either can be reduced to the other by multiplying its members by a constant, say r. Then the equations $Ax + By + C = 0$ and $rA'x + rB'y + rC' = 0$ are the same, term for term, so that $rA' = A$, $rB' = B$, and $rC' = C$. That is,

$$\frac{A}{A'} = \frac{B}{B'} = \frac{C}{C'} = r.$$

For example, if $mx + ny + 6 = 0$ and $4x - 2y + 3 = 0$ represent the same line, then $m/4 = n/-2 = 6/3$, and hence $m = 8$, $n = -4$.

The above proof assumes that A', B', and C' are not zero. If $A' = 0$, then $A = 0$, and both A and A' disappear from the equations. Similar remarks apply if $B' = 0$, or if $C' = 0$.

Exercise 26. Lines through Intersections

Find the equation of the line through the intersection of each of these pairs of lines and subject to the condition stated:

1. $3x - 2y = 13$ and $x + y - 6 = 0$, passing through $(2, -3)$.

2. $4x + y - 7 = 0$ and $3x - 2y = 10$, parallel to $x - 3y = 6$.

3. $3x + 5y - 13 = 0$ and $x + y - 1 = 0$, perpendicular to $7x - 5y = 10$.

4. Find q if the line $qx + 2y = 3$ passes through the intersection of the lines $4x + 3y = 7$ and $2x - y = 10$.

Since the equations $4x + 3y - 7 + k(2x - y - 10) = 0$ (§ 87) and $qx + 2y = 3$ represent the same line, § 89 may be used to find q.

5. Show that the lines $2x + y - 5 = 0$, $x - y + 2 = 0$, and $x + y = 4$ are concurrent.

That is, show that the third line, $x + y - 4 = 0$, is the same as the line $2x + y - 5 + k(x - y + 2) = 0$ for a certain value of k.

Exercise 27. Review

Find the equation of the line through the point $(-3, 4)$ and subject to the following conditions:

1. It is parallel to the line $5x + 4y = 6$.

2. It is perpendicular to the line through $(4, 1)$ and $(7, 3)$.

3. The x intercept is 10.

4. The sum of the intercepts is 12.

In Ex. 4, as in many of the other problems in this exercise, there is more than one line that satisfies the given conditions. The complete algebraic treatment of such a problem should be given, thus finding all the lines in each case.

5. The product of the intercepts is 50.

6. It is 2 units from the origin.

7. It is 5 units from the point $(12, 9)$.

8. It is as far as possible from the point $(10, 6)$.

9. It is equidistant from the points $(2, 2)$ and $(0, -6)$.

10. It passes through the intersection of the lines $x + y = 8$ and $4x - 3y = 12$.

11. The sum of the intercepts is equal to twice the excess of the y intercept over the x intercept.

Find the equation of the line with slope $-\frac{4}{3}$ and subject to the following conditions:

12. The x intercept is 6.

13. It forms with the axes a triangle whose perimeter is 24.

14. It is 6 units from the origin.

15. It is 4 units from the point $(10, 2)$.

16. It is equidistant from the points $(2, 7)$ and $(3, -8)$.

17. It is tangent to the circle $x^2 + y^2 - 4x - 8y = 5$.

That is, the distance from the line to the center of the circle is equal to the radius.

Find the equations of the following lines:

18. The line forming with the axes an isosceles triangle whose area is 60 square units.

19. The line through $(-1, 3)$ and having equal intercepts.

20. The perpendicular to the line $4x - y = 7$ at that point of the line whose abscissa is 1.

21. The line which has equal intercepts and is tangent to the circle with center $(4, -2)$ and radius 10.

22. The line parallel to the line $3x - y = 4$ and tangent to the circle $x^2 + y^2 = 9$.

23. The line midway between the parallel lines $3x - y = 4$ and $6x - 2y = 9$.

24. The line through the point $Q(4, 2)$ and cutting the x axis at A and the y axis at B so that $AQ : QB = 2 : 3$.

25. The line parallel to the parallel lines $3x + 4y = 10$ and $3x + 4y = 35$, and dividing the distance between them in the ratio $2 : 3$.

26. The line through the point $(0, 0)$ and forming with the lines $4x - y = 7$ and $x + 2y = 8$ an isosceles triangle.

A line through the origin is usually most conveniently represented by the equation $y = mx$.

27. The lines through the point $(4, 3)$ and tangent to the circle $x^2 + y^2 = 4$.

28. The lines tangent to the two circles $x^2 + y^2 = 16$ and $x^2 + y^2 - 2x - 8y = 19$.

29. The lines tangent to three circles whose centers are $(2, -3)$, $(5, -1)$, and $(-5, 1)$ and whose radii are proportional to 1, 2, and 3 respectively.

30. The line with slope $\frac{4}{3}$ and passing as near to the circle $x^2 + y^2 = 1$ as to the circle $x^2 + y^2 - 12x = -27$.

31. The diagonals of a parallelogram having as two sides $4x - 2y = 7$ and $3x + y = 6$ and as one vertex $(12, -3)$.

32. The hypotenuse of a right triangle is AB, where A is the point (3, 1) and B is the point (8, 7). If the abscissa of the vertex C of the right angle is 5, what is the ordinate of C?

33. The hypotenuse AB of an isosceles right triangle ABC joins (3, 1) and (0, 0). Find the coordinates of C.

34. The vertex C of the right angle of a right triangle ABC is (2, 3), and A is the point (4, -1). If the hypotenuse is parallel to the line $2x = y - 7$, find the equations of the three sides of the triangle.

35. Find the distance between the parallel lines $2x + y = 10$ and $2x + y = 15$.

36. Find the equations of the tangents from (7, 1) to the circle $x^2 + y^2 = 25$.

For any straight line prove that:

37. $b = -ma$. **38.** $m = -\cot\beta$. **39.** $a^2m^2 = p^2(1 + m^2)$.

Find the point which is:

40. Equidistant from the line $3y = 4x - 24$, the x axis, and the y axis.

41. On the x axis and 7 units from the line $4x + 3y = 15$.

42. In the second quadrant, equidistant from the axes and 2 units from the line $6x - 8y = -33$.

43. Equidistant from (4, 2) and (-2, 3) and also equidistant from the lines $2x + y = 10$ and $2x + 4y + 9 = 0$.

Given the points $A(2, 4)$, $B(8, 8)$, and $C(11, 3)$, draw the triangle ABC and find the following points:

44. M, the intersection of the medians.

45. L, the center of the circumscribed circle.

46. N, the intersection of the altitudes.

47. Show that the points L, M, and N, found in the three problems immediately preceding, are collinear. Show also that $NM : ML = 2 : 1$.

Given the points $A(0, 1)$, $B(0, 9)$, and $C(3, 5)$:

48. Find the point P such that the triangles BAP, ACP, and CBP are equivalent.

Because of the double sign ($\pm$) in the formula for the distance from a line to a point, there are four cases.

49. On the lines AB, BC, and CA lay off the segments 4, 5, and 10 respectively, and find P such that the triangles having P as a vertex and these segments as bases are equivalent.

50. Consider Ex. 49 for the case in which the three segments are a, b, and c respectively.

51. Let the three lines l_1, l_2, and l_3 be $2y = x$, $y = x$, and $y = 3x$ respectively. Then on l_1 locate the points $A(8, ?)$ and $A'(14, ?)$; on l_2 the points $B(7, ?)$ and $B'(9, ?)$; and on l_3 the points $C(2, ?)$ and $C'(5, ?)$.

52. The three pairs of corresponding sides of the triangles ABC and $A'B'C'$ in Ex. 51 meet in collinear points.

The corresponding sides are AB and $A'B'$, BC and $B'C'$, CA and $C'A'$, and so in all similar cases hereafter.

Taking any triangle ABC, letting the x axis lie along AB and the y axis bisect AB, and denoting the vertices by $A(-a, 0)$, $B(a, 0)$, $C(k, l)$, proceed as follows:

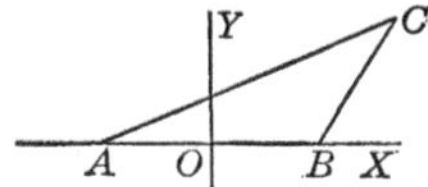

53. Find the common point P of the perpendicular bisectors of the sides.

54. Find the common point M of the medians.

55. Find the common point H of the altitudes.

56. Using the notation of Exs. 53–55, prove that P, M, and H are collinear.

57. Show that the points P, M, and H referred to in Exs. 53–56 are such that $PM : MH = 1 : 2$.

It is not necessary to find the lengths PM and MH.

If the parallel lines $Ax + By + C = 0$ and $Ax + By + C' = 0$ are represented by $L = 0$ and $L' = 0$ respectively, show that:

58. $L + kL' = 0$ represents a line parallel to them.

59. $L + L' = 0$ represents a line midway between them.

60. $L - L' = 0$ represents no points in the plane.

61. If $L = 0$ and $L' = 0$ represent the normal equations of any two lines, then $L + L' = 0$ and $L - L' = 0$ represent lines which are perpendicular to each other.

It should be noted that the symbols L and L' in Ex. 61 do not denote the same expressions as in Exs. 58–60.

62. In the system of lines $(k+1)x + (k-1)y = k^2 - 1$ the difference between the intercepts is the same for all the lines.

63. Draw the lines of the system in Ex. 62 for the following values of k: $-5, -4, -3, -2, -1, 0, 1, 2, 3, 4, 5$. Describe the arrangement of the system.

It will be found best to use coordinate paper with ten squares to the inch. Extend every line to the margins.

64. The lines of the system $kx + 4y = k$ are concurrent.

Find the common point of any two lines of the system and show that this point is on each of the other lines.

65. If r is the radius of a circle with center at $(0, 0)$, all lines of the system $y = mx \pm r\sqrt{1 + m^2}$ are tangent to the circle.

66. All lines represented by the equation

$$(k + h)x + (k - h)y = 10\sqrt{k^2 + h^2}$$

are tangent to the circle $x^2 + y^2 = 50$.

67. Rays from a point of light at $A(10, 0)$ are reflected from the y axis. Find an equation which represents all the reflected rays and prove that all these reflected rays, when produced through the y axis to the left, pass through $(-10, 0)$.

The direct ray and the reflected ray make equal angles with OY.

From A draw a ray to any point $B(0, k)$ of the y axis. Then draw the reflected ray, find its slope in terms of k, and write its equation.

If $4x - 3y = 12$ represents the line l_1 and $12x + 5y = 0$ represents the line l_2, find the equation of the locus of P under the following conditions:

68. P is twice as far from l_1 as from l_2.

69. P is equidistant from l_1 and l_2.

70. P is as far from O as from l_1.

71. P is the vertex of a triangle of area 20 and of base 8, the other two vertices lying on l_1.

72. $AB = 10$ and $CD = 13$, AB is a segment of l_1 and CD is a segment of l_2, and P moves so that the triangles PAB and PCD have equal areas.

73. Find the equations of the bisectors of the angles formed by the lines $Ax + By + C = 0$ and $A'x + B'y + C' = 0$.

74. Using oblique axes with any angle, find the equation of the line through the points $P_1(x_1, y_1)$ and $P_2(x_2, y_2)$.

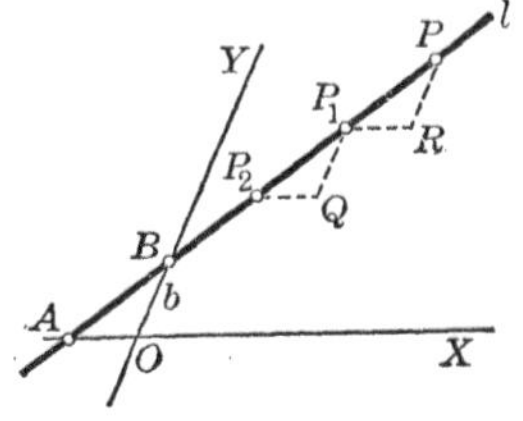

Draw P_2Q, P_1Q, P_1R, and PR parallel to the axes, $P(x, y)$ being any point on the line l. The required equation is then found from the condition $RP/P_1R = QP_1/P_2Q$. But $RP = y - y_1$, and similarly for other values. It will be seen that the result is the same as in rectangular coordinates.

75. Using oblique axes, find the equation of a line in terms of its intercepts.

Employing the result of Ex. 74, use the same method as the one employed for rectangular coordinates (§ 77).

76. In the figure of Ex. 74, denote the constant ratio $QP_1 : P_2Q$ or $RP : P_1R$ by m, and find the equation of the line l when m and a point $P_1(x_1, y_1)$ are given.

77. As in Exs. 74 and 76, find the equation of l when m and b are given.

The equations found in Exs. 74–77 are identical in form with the corresponding equations in rectangular coordinates; but m is not equal to $\tan \alpha$, that is, m is not the slope of l.

If OD and OB are any two lines, and AB and AC any other two lines, the oblique axes being taken as shown in the figure, and the points A, B, and E being $(a, 0)$, $(0, b)$, and (e, f) respectively, find the following:

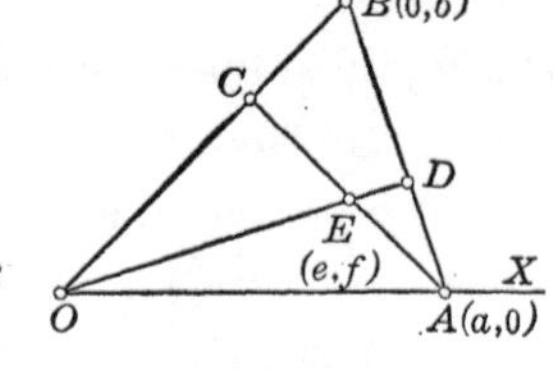

78. The equations of OB and AC.

79. The coordinates of C.

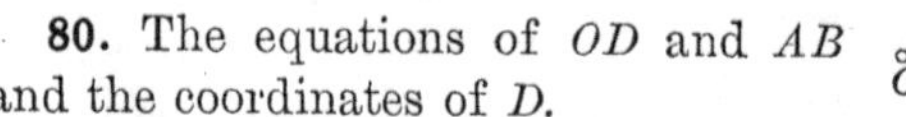

80. The equations of OD and AB and the coordinates of D.

81. From the coordinates of C and D as found in Exs. 79 and 80 find the equation of CD, and find the coordinates of the point Q in which CD cuts the x axis.

82. Find the coordinates of the point R in which BE cuts OA.

83. Show that the points R and Q divide OA internally and externally, respectively, in the same ratio.

84. For all lines in the plane prove that $\frac{1}{a^2} + \frac{1}{b^2} = \frac{1}{p^2}$, where a, b, and p have their usual meanings.

85. Since in any given length the number of centimeters c is proportional to the number of inches i, it is evident that $c = ki$, where k is a constant. Given that 10 in. = 25.4 cm., find the value of k and plot the graph of the equation $c = ki$. From the graph estimate the number of inches in 20 cm.

If the graph is accurately drawn on a reasonably large scale, any desired number of inches may be converted into centimeters at a glance, and vice versa. Such a graph is called a *conversion graph*.

86. Given that 2 cu. ft. = 15 gal., find the conversion formula for gallons and cubic feet, and draw the graph.

87. Two variables x and y are related by the linear formula $y = ax + b$. If, by substituting for x any value, say x', we obtain for y the value y', and if, by substituting for x a changed value $x' + h$, we obtain for y the changed value $y' + k$, prove that $k = ah$. That is to say, prove that the change in y is proportional to the change in x.

88. A wheel 3 ft. in diameter makes 20 rev./sec. A brake is then applied and the velocity of the rim decreases F ft./sec., F being a constant. If the wheel comes to rest in 4 sec., find a formula for the velocity v of the rim in ft./sec. at the end of t seconds after the brake is applied, and draw the graph.

The symbol "rev./sec." is read "revolutions per second," and similarly in the other cases.

89. The height h of the mercury in a barometer falls practically 0.11 in. for each 100 ft. that the barometer is carried above sea level, up to a height of 2000 ft. An airplane ascends from a point 500 ft. above sea level, the barometer at that place reading 29.56 in. Find in terms of h the formula for the height H of the airplane above sea level, assuming that H does not exceed 2000 ft. and that there is no disturbance in the weather conditions. Draw the graph.

It is evident that $h = c - 0.11\,H/100$, where h and c are measured in inches and c, the barometric reading at sea level, is to be found.

90. If E is the effort required to raise a weight W with a pulley block, E and W are connected by a linear relation. If it is given that $W = 415$ when $E = 100$, and $W = 863$ when $E = 212$, find the linear relation and draw the graph.

91. In testing a certain type of crane it was observed that the pull P required to lift the weight W was as follows:

$P =$	48	65	83.5	106.8	129	145
$W =$	800	1200	1600	2000	2400	2800

Locate points with these pairs of values and note that they lie approximately on a straight line. This suggests a linear relation between P and W. Draw the straight line which seems to fit the data best and find its equation, measuring the coordinates of two of its points for this purpose.

The two points should be taken some distance apart. Different units may be used on the two axes.

92. From the equation found in Ex. 91 compute the values of W for the values of P given in the table and compare the results with the given values of W.

Some values of W found from the equation should be more and others less than the values given in the table. Add together such of the differences as represent excesses over the true values and also add such of the differences as represent deficiencies and compare the totals. If these totals are about equal, leave the line as it is; but if one of them differs considerably from the other, move the line accordingly, find the equation again, and then test it as before.

93. In a study of the friction between two oak surfaces it was found that the following table shows the pull P which is required to give a slow motion to the weight W:

$P =$	5	12	19.4	26.2	34	39.5	48
$W =$	2	4	6	8	10	12	14

Find the relation between P and W as in Ex. 91.

94. Test the equation of Ex. 93 as in Ex. 92.

95. The condition in determinant form that the three lines $Ax + By + C = 0$, $A'x + B'y + C' = 0$, and $A''x + B''y + C'' = 0$ shall be concurrent is that

$$\begin{vmatrix} A & B & C \\ A' & B' & C' \\ A'' & B'' & C'' \end{vmatrix} = 0.$$

Ex. 95 should be omitted by those who have not studied determinants.

96. From $P(4, 6)$ and $Q(7, 1)$ draw the perpendiculars PR and QS to the line $5x - 3y = 36$, and find the length of RS.

RS is called the *projection* of the segment PQ upon the line. For a simple solution, see Ex. 97.

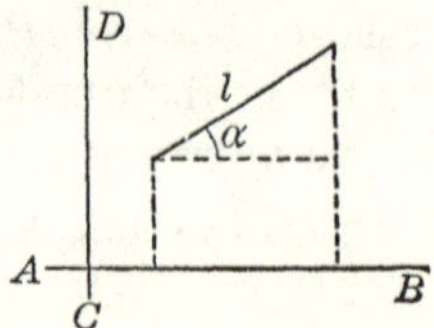

97. Show that the projection of the line l upon the line AB is equal to $l \cos \alpha$, and then find the projection of l upon a line CD which is perpendicular to AB.

CHAPTER V

THE CIRCLE

PROBLEM. EQUATION OF THE CIRCLE

90. *To find the equation of the circle with given center and given radius.*

Solution. Let $C(a, b)$ be the center and r the radius of the circle. Then if $P(x, y)$ is any point on the circle, the circle is defined by the condition that

$$CP = r,$$

and since $\sqrt{(x-a)^2+(y-b)^2} = CP,$

we have $\sqrt{(x-a)^2+(y-b)^2} = r,$

or
$$(x-a)^2 + (y-b)^2 = r^2. \tag{1}$$

This is the desired equation. It may be written

$$x^2 + y^2 - 2\,ax - 2\,by + a^2 + b^2 - r^2 = 0,$$

or
$$x^2 + y^2 - 2\,ax - 2\,by + c = 0, \tag{2}$$

in which c stands for $a^2 + b^2 - r^2$.

If the center of the circle is the origin, then $a = 0$, $b = 0$, and the equation is

$$x^2 + y^2 = r^2.$$

It must not be inferred that every equation which has one of the above forms represents a real circle. For example, there are no real points (x, y) that satisfy the equation

$$(x-2)^2 + (y+3)^2 = -20,$$

because $(x-2)^2 + (y+3)^2$, being the sum of two squares, cannot be negative.

THEOREM. CONVERSE OF § 90

91. *An equation of the form* $x^2 + y^2 - 2ax - 2by + c = 0$ *represents a real circle except when* $a^2 + b^2 - c$ *is negative.*

Proof. Writing the given equation in the form

$$x^2 - 2ax + y^2 - 2by = -c$$

and completing both squares in the first member, we have

$$x^2 - 2ax + a^2 + y^2 - 2by + b^2 = a^2 + b^2 - c,$$

or

$$(x-a)^2 + (y-b)^2 = a^2 + b^2 - c.$$

Letting $a^2 + b^2 - c = r^2$, we have

$$(x-a)^2 + (y-b)^2 = r^2,$$

an equation which evidently (§ 90) represents a real circle with center (a, b) and radius $r = \sqrt{a^2 + b^2 - c}$ except when $a^2 + b^2 - c$ is negative.

When $a^2 + b^2 - c = 0$, we see at once that $r = 0$. Then the equation $(x-a)^2 + (y-b)^2 = 0$, since it is satisfied by no values except $x = a$ and $y = b$, represents only one point, (a, b). This point is called a *point circle.*

When $a^2 + b^2 - c$ is negative, say equal to $-k$, we say that the equation $(x-a)^2 + (y-b)^2 = -k$ represents an *imaginary circle.*

92. COROLLARY. *An equation of the second degree in* x *and* y *represents a circle if and only if it has no term in* xy *and the coefficients of* x^2 *and* y^2 *are equal.*

For such an equation, say $Ax^2 + Ay^2 + Bx + Cy + D = 0$, may be written

$$x^2 + y^2 + \frac{B}{A}x + \frac{C}{A}y + \frac{D}{A} = 0,$$

which is in the form $x^2 + y^2 - 2ax - 2by + c = 0$.

It will be observed that many equations of the second degree in x and y do not represent circles, as, for example, the equation $9x^2 + 25y^2 = 225$, which was discussed in § 42.

Exercise 28. Center and Radius

Find the center and the radius of each of the following circles and draw the figure:

1. $x^2 + y^2 - 8x + 4y = 5$.
2. $x^2 + y^2 - 12x - 2y = 12$.
3. $x^2 + y^2 + 8x + 6y = 0$.
4. $2x^2 + 2y^2 - 8x + 10y = 11\frac{1}{2}$.
5. $x^2 + y^2 - 6x = 0$.
6. $x^2 + y^2 - 6x = 16$.
7. $x^2 + y^2 + 8y = 0$.
8. $x^2 + y^2 + 2x + 2 = 2y$.

9. The circle $x^2 + y^2 - 2ax = 0$ is tangent to the y axis at the origin.

10. The circle $x^2 + y^2 + 8x - 4y + 16 = 0$ is tangent to the x axis.

11. Given that the circle $(x - 2)^2 + (y - 5)^2 = r^2$ passes through $(10, -1)$, find the value of r.

12. Find the area of the square circumscribed about the circle $x^2 + y^2 + 4x + 4y = 8$.

13. Draw the two circles $x^2 + y^2 - 4x - 6y + 9 = 0$ and $x^2 + y^2 + 12x + 6y - 19 = 0$, and prove that they are tangent.

Show that the line joining the centers is equal to the sum of the radii.

14. The circle $x^2 + y^2 - 18x + 45 = 0$ is tangent to the line $y = \frac{4}{3}x - 2$.

15. Show that the equation

$$x^2 + y^2 + x + k(x^2 + y^2 - 2x + y - 1) = 0$$

represents a circle, and find the center and the radius.

Draw the circles described below, and find their equations:

16. Center $(-1, 2)$, radius 6.

17. Center $(4, 0)$, tangent to the line $x = 8$.

18. Center $(3, 4)$, tangent to the line $8y = 15 - 6x$.

19. Passing through the points $(4, 0)$, $(0, -8)$, and $(0, 0)$.

20. Tangent to the lines $x = 6$, $x = 12$, and $y = 8$.

21. Radius 10, x intercepts 0 and 12.

93. Three Essential Constants. The equation of the circle involves three essential constants. If the equation is in the form $(x-a)^2+(y-b)^2=r^2$, these constants are a, b, r; if it is in the form $x^2+y^2-2ax-2by+c=0$, the constants are a, b, c.

In either case it is evident that three constants are necessary and sufficient to fix the circle in size and in position.

To find the equation of the circle determined by three conditions we choose either of the two standard forms, $(x-a)^2+(y-b)^2=r^2$ or $x^2+y^2-2ax-2by+c=0$. The problem then reduces to expressing the three conditions in terms of the three constants of the form selected.

94. Illustrative Examples. **1.** Find the equation of the circle through the points $A(4, -2)$, $B(6, 1)$, $C(-1, 3)$.

Let $x^2+y^2-2ax-2by+c=0$ represent the circle. Then, since A is on the circle, its coordinates, 4 and -2, satisfy the equation.

That is, $$4^2+(-2)^2-2a\cdot 4-2b(-2)+c=0,$$

whence $$8a-4b-c=20.$$

Similarly, for B, $$12a+2b-c=37,$$

and for C, $$2a-6b+c=-10.$$

Solving, we have $a=\frac{23}{10}$, $b=\frac{13}{10}$, $c=-\frac{34}{5}$, and the equation is

$$x^2+y^2-\tfrac{23}{5}x-\tfrac{13}{5}y-\tfrac{34}{5}=0,$$

or $$5x^2+5y^2-23x-13y-34=0.$$

2. Find the equation of the circle through $(2, -1)$, tangent to $x+y=1$ and having its center on $y=-2x$.

If we choose the form $(x-a)^2+(y-b)^2=r^2$, the first condition is $(2-a)^2+(-1-b)^2=r^2$, or $a^2+b^2-4a+2b+5=r^2$.

Since the distance from the line $x+y=1$ to the center (a, b) is r, the second condition is $(a+b-1)/(\pm\sqrt{2})=r$.

Since (a, b) is on the line $y=-2x$, the third condition is $b=-2a$.

Solving for a, b, and r^2, we have $a=1$, $b=-2$, $r^2=2$, and the required equation is $(x-1)^2+(y+2)^2=2$.

Exercise 29. Equations of Circles

Draw circles passing through the following sets of points, and find the equation of each circle:

1. (4, 2), (5, – 1), (– 2, 4).
2. (6, – 3), (4, – 2), (0, 4).
3. (0, 4), (0, – 4), (6, 0).
4. (5, 1), (3, 2), (3, 1).

Draw circles subject to the following conditions, and find the equation of each circle:

5. Having x intercepts 6 and 10, and one y intercept 8.
6. Having x intercepts – 4 and 2, and radius 5.
7. Passing through (0, 0) and (–1, 1), and having radius 5.
8. Having a diameter joining (6, 3) and (– 2, – 5).
9. Tangent to the x axis at (6, 0) and tangent to the y axis.
10. Tangent to both axes and passing through (2, 1).
11. Tangent to both axes and to the line $2x + y = 6 + \sqrt{20}$.
12. Tangent to the lines $x = 6$ and $x = 10$, and passing through the point (8, 3).
13. Having one y intercept 10 and tangent to the x axis at the point (– 5, 0).
14. Passing through (0, 0) and the common points of the circles $x^2 + y^2 = 25$ and $x^2 + y^2 - 4x + 2y = 15$.
15. Inscribed in the triangle whose sides are the lines $4x - 3y = 12$, $8x + 6y = 36$, and $12x - 5y = 36$.
16. Passing through (0, 0) and cutting a chord $5\sqrt{2}$ units in length from each of the lines $x - y = 0$ and $x + y = 0$.
17. Having for a diameter that segment of the line $y = mx$ which is intercepted by the circle $x^2 + y^2 - 2ax = 0$.
18. Tangent to the lines $4x - 3y = 10$ and $6x + 8y = 35$, and having radius 10.
19. Tangent to the circle $x^2 + y^2 + 4x - 4y = 1$, and having its center at the point (4, 10).

PROBLEM. EQUATION OF A TANGENT

95. *To find the equation of the tangent to the circle* $x^2 + y^2 = r^2$ *at any point* $P_1(x_1, y_1)$ *on the circle.*

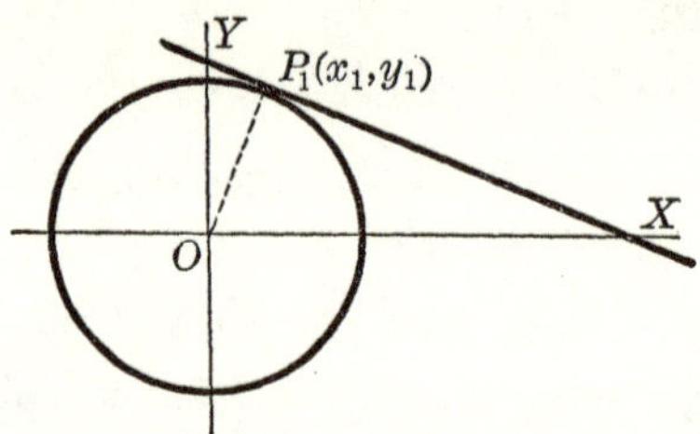

Solution. The center O is at the origin. § 90

The slope of the radius OP_1 is $\frac{y_1}{x_1}$. § 21

Since the tangent at P_1 is perpendicular to OP_1, the slope of the tangent at P_1 is $-\frac{x_1}{y_1}$. § 24

Therefore the equation of the tangent at P_1 is

$$y - y_1 = -\frac{x_1}{y_1}(x - x_1), \qquad \S\ 69$$

or

$$x_1x + y_1y = x_1^2 + y_1^2.$$

This is the required equation. It may be simplified in the following manner:

Since the point (x_1, y_1) is on the circle, we have

$$x_1^2 + y_1^2 = r^2,$$

and hence the equation of the tangent as deduced above may be more simply written

$$\boldsymbol{x_1x + y_1y = r^2.}$$

It will be noticed that the equation of the tangent is simply the equation of the circle $x^2 + y^2 = r^2$ with x_1x written for x^2 and y_1y written for y^2.

PROBLEM. TANGENT AT A GIVEN POINT

96. *To find the equation of the tangent to the circle* $x^2+y^2-2ax-2by+c=0$ *at any point* $P_1(x_1, y_1)$ *on the circle.*

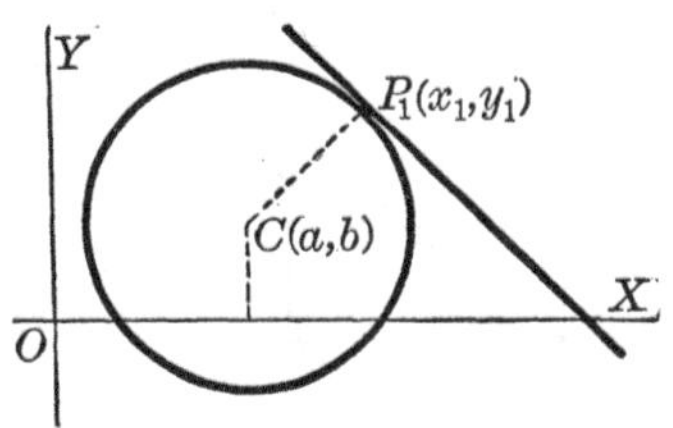

Solution. The center C of the circle is (a, b). § 90

The slope of the radius CP_1 is $\frac{y_1-b}{x_1-a}$. § 21

Hence the slope of the tangent at P_1 is $-\frac{x_1-a}{y_1-b}$. § 24

Therefore the equation of the tangent at P_1 is

$$y-y_1=-\frac{x_1-a}{y_1-b}(x-x_1), \quad \text{§ 69}$$

or $$x_1x+y_1y-ax-by-(x_1^2+y_1^2-ax_1-by_1)=0.$$

This equation may be simplified in the following manner:

Since the point (x_1, y_1) is on the circle, we have

$$x_1^2+y_1^2-2ax_1-2by_1+c=0, \quad \text{§ 90}$$

whence $$x_1^2+y_1^2-ax_1-by_1=ax_1+by_1-c.$$

Therefore the equation of the tangent may be written

$$x_1x+y_1y-ax-by-(ax_1+by_1-c)=0,$$

or in simpler form

$$\mathbf{x_1x+y_1y-a(x+x_1)-b(y+y_1)+c=0.}$$

As in § 95, it will be noticed that the equation of the tangent is simply the equation of the circle $x^2+y^2-2ax-2by+c=0$ with x_1x written for x^2, y_1y for y^2, $x+x_1$ for $2x$, and $y+y_1$ for $2y$.

PROBLEM. TANGENTS WITH A GIVEN SLOPE

97. *To find the equations of the tangents to the circle* $x^2 + y^2 = r^2$ *having a given slope* m.

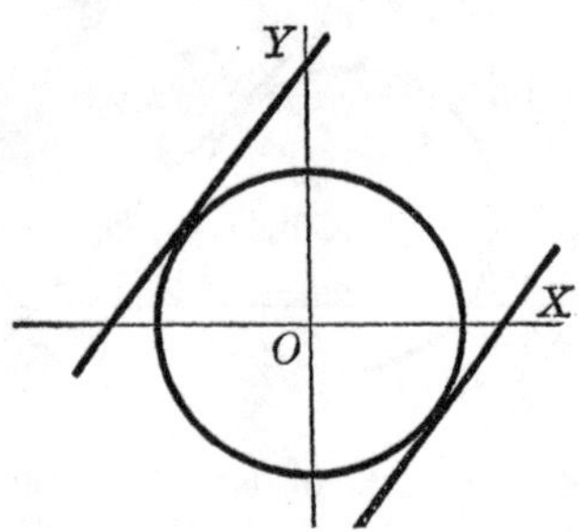

Solution. If we let $y = mx + c$ represent any line of slope m, the problem reduces to finding the value of c for which the line $y = mx + c$ is tangent to the circle $x^2 + y^2 = r^2$.

To find the points in which the line cuts the circle, we solve the equations as simultaneous. Substituting, we have

$$x^2 + (mx + c)^2 = r^2,$$

or

$$(1 + m^2)x^2 + 2\,mcx + c^2 - r^2 = 0.$$

The two roots of this quadratic in x are the abscissas of the common points of the line and the circle; but in order that the line shall be tangent, these points must coincide and thus have the same abscissa, and hence the roots of this quadratic in x must be equal. Since the condition that the roots of any quadratic $Ax^2 + Bx + C = 0$ shall be equal is that $B^2 - 4\,AC = 0$, we must have

$$4\,m^2c^2 - 4(1 + m^2)(c^2 - r^2) = 0,$$

whence,

$$c = \pm r\sqrt{1 + m^2}.$$

Therefore the required equations of the tangents are

$$\boldsymbol{y = mx \pm r\sqrt{1 + m^2}}.$$

98. Normal. The line which is perpendicular to a tangent to a curve at the point of contact is called the *normal* to the curve at that point.

99. Angle between Two Circles. When two circles intersect and a tangent is drawn to each circle at either point of intersection, the angle between these tangents is called the *angle between the circles.*

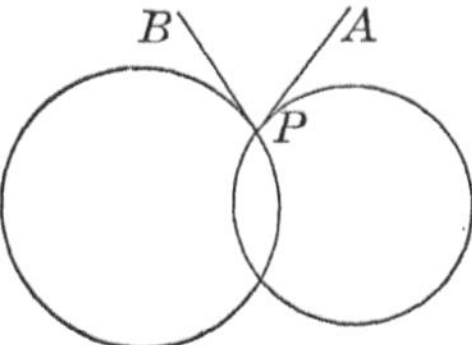

In the figure the angle between the two circles is APB.

100. Illustrative Examples. 1. Find the equations of the tangent and the normal to the circle $x^2 + y^2 = 29$ at the point $(5, -2)$.

Since $5^2 + (-2)^2 = 29$, the point $(5, -2)$ is on the circle, and hence the equation of the tangent at this point is $5x - 2y = 29$ (§ 95).

Since the normal passes through $(5, -2)$ and is perpendicular to the tangent, its equation (§ 69) is $y + 2 = -\frac{2}{5}(x - 5)$, or $2x + 5y = 0$.

2. Find the equations of the tangents to the circle $x^2 + y^2 = 13$ which pass through the point $(-4, 7)$.

Since $(-4)^2 + 7^2$ is not equal to 13, the point $(-4, 7)$ is not on the circle and therefore the point of contact is unknown. In such cases it is often better to use the equation $y = mx \pm r\sqrt{1 + m^2}$.

In this example $r = \sqrt{13}$, and m is to be chosen so that the tangent $y = mx \pm \sqrt{13}\sqrt{1 + m^2}$ passes through $(-4, 7)$; therefore $7 = -4m \pm \sqrt{13}\sqrt{1 + m^2}$. Solving this equation, $m = -\frac{2}{3}$ or -18.

If $m = -\frac{2}{3}$, the tangents are $y = -\frac{2}{3}x \pm \sqrt{13}\sqrt{1 + \frac{4}{9}}$ (§ 97), or $2x + 3y = \pm 13$, but only $2x + 3y = 13$ passes through $(-4, 7)$.

If $m = -18$, the tangents are $y = -18x \pm \sqrt{13}\ \sqrt{1 + 324}$ or $18x + y = \pm 65$, but only $18x + y = -65$ passes through $(-4, 7)$.

3. If the line $4x - 3y = 50$ is tangent to the circle $x^2 + y^2 = 100$, find the point of contact.

If (x_1, y_1) is the point of contact, $x_1x + y_1y = 100$ is the tangent (§ 95). But the tangent is $4x - 3y = 50$. Hence $x_1 : 4 = y_1 : -3 = 100 : 50$, by § 89. Then $x_1 = 8$, $y_1 = -6$, and the point (x_1, y_1) is $(8, -6)$.

Exercise 30. Tangents and Normals

In each of these cases show that P_1 is on the given circle, and find the equation of the tangent and also of the normal at P_1:

1. $x^2 + y^2 = 25$; $P_1(3, 4)$.
2. $x^2 + y^2 = 29$; $P_1(2, -5)$.
3. $x^2 + y^2 = 34$; $P_1(-5, 3)$.
4. $x^2 + y^2 = 1$; $P_1(\frac{1}{2}, -\frac{1}{2}\sqrt{3})$.
5. $x^2 + y^2 = 20$; $P_1(-4, -2)$.
6. $x^2 + y^2 = 37$; $P_1(-1, -6)$.
7. $x^2 + y^2 - 6x + 2y = 0$; $P_1(2, 2)$.
8. $x^2 + y^2 + 4x - 7y - 11 = 0$; $P_1(3, 2)$.

Find the equations of the tangents to the following circles under the stated conditions, and find the points of contact:

9. $x^2 + y^2 = 25$; the slope of the tangent is $\frac{3}{4}$.
10. $x^2 + y^2 = 49$; the slope of the tangent is $-\frac{12}{5}$.
11. $x^2 + y^2 = 36$; the tangent is parallel to $4x = 3y$.
12. $x^2 + y^2 = 13$; the tangent is perpendicular to $x = \frac{2}{3}y$.
13. $x^2 + y^2 = 104$; the point $(-8, 12)$ is on the tangent.
14. $x^2 + y^2 = 64$; the tangent passes through $(8, 4)$.

Since $y = mx + \sqrt{64}\sqrt{1 + m^2}$ must be satisfied by $(8, 4)$ we have the equation $4 = 8m + 8\sqrt{1 + m^2}$, whence $1 - 4m + 4m^2 = 4 + 4m^2$, or $0 \cdot m^2 + 4m + 3 = 0$. This may be considered as a special quadratic, and its solution is considered in § 6 on page 283 of the Supplement.

15. In this figure, if PF represents the direction and intensity of a force applied to a wheel at P, PF is the resultant of two other forces, a *tangential component* PT which turns the wheel, and a *normal component* PN which produces no turning. Then PF is the diagonal of the parallelogram of forces of which PT and PN are sides. If, now, it is given that P is $(4, 3)$, F is $(\frac{56}{5}, \frac{9}{10})$, and OA is 5, find the values of PT and PN.

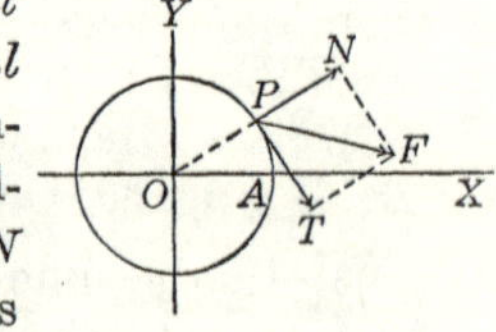

16. Find the angle between the circles $x^2 + y^2 - 34 = 0$ and $x^2 + y^2 - 5x + 10y + 1 = 0$.

Theorem. System of Circles through Two Points

101. *Denoting the two expressions* $x^2 + y^2 - 2ax - 2by + c$ *and* $x^2 + y^2 - 2a'x - 2b'y + c'$ *by* S *and* S' *respectively, the equation* $S + kS' = 0$, *k being an arbitrary constant, represents the system of circles through the two points of intersection of the circles* $S = 0$ *and* $S' = 0$.

Proof. Writing $S + kS' = 0$ in full, we have

$$x^2 + y^2 - 2ax - 2by + c + k(x^2 + y^2 - 2a'x - 2b'y + c') = 0,$$

or $(1+k)x^2 + (1+k)y^2 - 2(a + a'k)x - 2(b + b'k)y + c + c'k = 0$, which (§ 92) represents a circle for every value of k, with the exception of the value $k = -1$.

Since each pair of values that satisfies both $S = 0$ and $S' = 0$ also satisfies $S + kS' = 0$, the circle $S + kS' = 0$ passes through the common points of the circles $S = 0$ and $S' = 0$.

And since the substitution of x_1 for x and of y_1 for y in the equation $S + kS' = 0$ leaves k the only unknown, we see that k can be so determined that the circle $S + kS' = 0$, which passes through the common points Q and R of the two circles, passes through any third point (x_1, y_1).

Hence $S + kS' = 0$ represents all circles through Q and R.

For example, the circle passing through $(2, -1)$ and the common points of the circles $x^2 + y^2 - x + 2y = 3$ and $x^2 + y^2 - 6x = 4$ is $x^2 + y^2 - x + 2y - 3 + k(x^2 + y^2 - 6x - 4) = 0$, where k is to be so chosen that the equation is satisfied by $x = 2$ and $y = -1$; that is, $k = -\frac{2}{11}$. Hence the required equation is $9x^2 + 9y^2 + x + 22y = 25$.

102. Radical Axis. If $k = -1$, the equation $S + kS' = 0$ becomes $S - S' = 0$, or $2(a - a')x + 2(b - b')y + c' - c = 0$. This equation, therefore, represents the straight line through the common points of the circles $S = 0$ and $S' = 0$. Such a line is called the *radical axis* of the two circles.

PROBLEM. LENGTH OF A TANGENT

103. *To find the length of the tangent from any external point $P_1(x_1, y_1)$ to the circle $(x-a)^2+(y-b)^2=r^2$.*

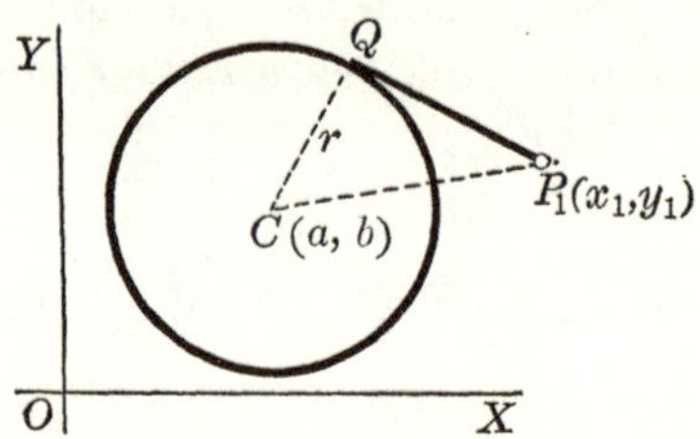

Solution. Let C (a, b) be the center of the circle and Q the point of tangency. Then in the right triangle CP_1Q the side P_1Q is the tangent the length of which is required.

Since $\overline{P_1Q}^2 = \overline{CP_1}^2 - r^2$

$= (x_1-a)^2+(y_1-b)^2-r^2,$ § 17

we have $$P_1Q = \sqrt{(x_1-a)^2+(y_1-b)^2-r^2}.$$

That is, *the length of the tangent from an external point (x_1, y_1) to any circle may be found by transposing to the first member all the terms of the equation of the circle, substituting x_1 for x and y_1 for y, and taking the square root of the result.*

Exercise 31. Two Circles and Lengths of Tangents

1. The radical axis of any two circles is perpendicular to the line of centers.

2. The radical axis of any two real circles is real, even if the circles do not intersect in real points.

3. The tangents to the circles $S=0$ and $S'=0$ from any point on the radical axis are equal.

4. The three radical axes determined by three circles meet in a point.

Exercise 32. Review

1. In the circle $x^2 + y^2 - 4x + 6y = 12$, find the x intercepts and the sum, product, and difference of these intercepts.

2. Find the sum, difference, and product of the x intercepts and of the y intercepts of the circle $x^2 + y^2 + 2x - 13y + 40 = 0$.

Given any circle $x^2 + y^2 - 2ax - 2by + c = 0$, *prove that:*

3. The difference of the x intercepts is $2\sqrt{a^2 - c}$.

4. The sum of the x intercepts is $2a$ and the product is c.

5. The sum of the y intercepts is $2b$ and the product is c.

6. The difference of the squares of the chords cut by the circle from the axes is $4(a^2 - b^2)$.

Find the equation of each of the following circles:

7. Tangent to OY at the point (0, 6) and cutting from OX a chord 16 units in length.

Notice that the chord is equal to the difference of the x intercepts, and then use the result of Ex. 3.

8. With center (– 3, 5), the product of the four intercepts being equal to 225.

9. Tangent to both axes and having the area included between the circle and the axes equal to $16(4 - \pi)$.

10. Passing through (4, 2) and (– 1, 3) and having the sum of the four intercepts equal to 14.

After the equation of the circle has been found, draw the circle, find the four intercepts, and note that two of these intercepts are imaginary. How much do these imaginary intercepts contribute to the given sum?

11. Show that the square of the tangent from the origin to the circle $x^2 + y^2 - 2ax - 2by + c = 0$ is equal to c, and therefore show that the origin is outside the circle when $c > 0$, on the circle when $c = 0$, and inside the circle when $c < 0$.

12. A ray of light from (10, 2) strikes the circle $x^2 + y^2 = 25$ at (4, 3). Find the equation of the reflected ray.

13. Find the equation of the circle having (a, b) as center and passing through the origin.

14. The tangent to the circle $x^2 + y^2 - 2\,ax - 2\,by = 0$ at the origin is $ax + by = 0$.

15. Show that the tangents drawn from the origin to the circle $(x - a)^2 + (y - b)^2 = r^2$ are also tangent to the circle $(x - ka)^2 + (y - kb)^2 = (kr)^2$, where k is any constant.

16. The circles $x^2 + y^2 = 36$ and $x^2 + y^2 - 24\,x = 108$ intersect at right angles.

Such circles are said to be *orthogonal.*

17. The condition that the circle $x^2 + y^2 + Dx + Ey + F = 0$ shall be tangent to the x axis is that $D^2 = 4\,F$.

18. Find the condition that $x^2 + y^2 + Dx + Ey + F = 0$ and $x^2 + y^2 + D'x + E'y + F = 0$ shall be tangent circles.

19. Find the equations of the three circles with centers (2, 5), (5, 1), and (8, 5), each circle being tangent to both the others.

20. Find the points of intersection of the line $3\,x - y = 3$ with the circle $x^2 + y^2 + x - 4\,y - 3 = 0$.

21. Find the points of intersection of the line $2\,x + y + 3 = 0$ with the circle $x^2 + y^2 - 4\,x - 6\,y = 7$.

22. Find the points of intersection of the x axis with the circle $x^2 + y^2 - 2\,x + 4\,y = 8$.

23. Find the condition that the line $lx + my + n = 0$ shall be tangent to the circle $x^2 + y^2 = r^2$.

24. Find the equation of the circle with radius 10 and tangent to the line $3\,y = 4\,x + 3$ at the point (3, 5).

25. Given that the circle $x^2 + y^2 - 6\,x + ky + l = 0$ is tangent to the line $4\,y = 3\,x - 8$ at the point (4, 1), find k and l.

The results found in §§ 96 and 89 may be employed to advantage.

26. Find the equation of the chord through the two points of contact of the tangents from $P(-1, 7)$ to the circle $x^2 + y^2 = 25$.

Such a chord is called the *chord of contact.*

27. The equation of the chord of contact of the tangents from (x_1, y_1) to the circle $x^2 + y^2 = r^2$ is $x_1x + y_1y = r^2$.

How is this fact related to the fact that when (x_1, y_1) is on the circle the equation of the tangent at (x_1, y_1) is $x_1x + y_1y = r^2$? The corresponding problem for the parabola is solved in § 215.

28. A wheel with radius $\frac{1}{2}\sqrt{41}$ ft. is driven by steam, the pressure being transmitted by the rod AB, which is 10 ft. long. If the thrust on AB is 25,000 lb. when B is at $(2, \frac{5}{2})$, find the corresponding tangential and normal components of the thrust.

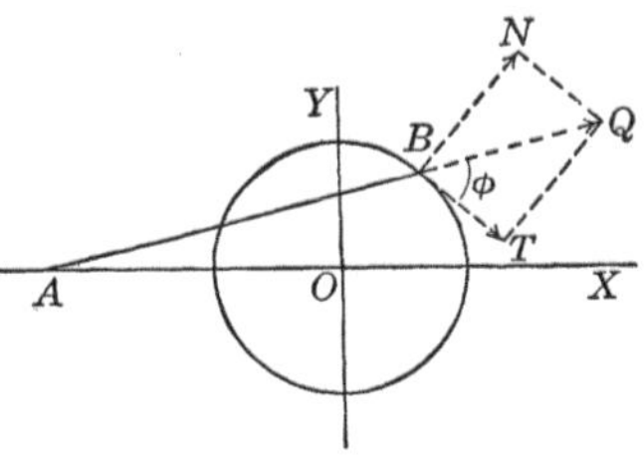

For the explanation of the technical terms, see Ex. 15, page 100. It will be found convenient to take the origin at the center of the circle. If the segment BQ represents the thrust of 25,000 lb., BT and BN represent the required components. Since $BT = BQ \cos\phi$, the problem reduces to finding ϕ by means of the slopes of BT and BQ.

29. Two forces, $F = 20$ lb. and $F' = 30$ lb., are applied at $P(12, 5)$, a point on the circle $x^2 + y^2 = 169$, F acting along a line with slope $\frac{1}{4}$ and F' along a line with slope 2. Find the sum of the normal components of F and F'.

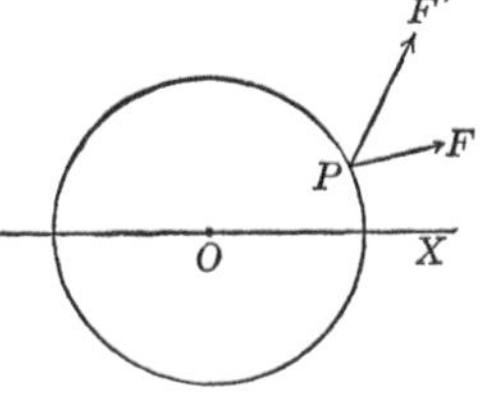

30. If a wheel with radius 5 ft. rotates so that a point P on the rim has a speed of 10 ft./sec., find the vertical and the horizontal components of the speed of P when P is at the point $(-4, 3)$.

In the figure, if $PQ = 10$, the problem is to find PV and PH.

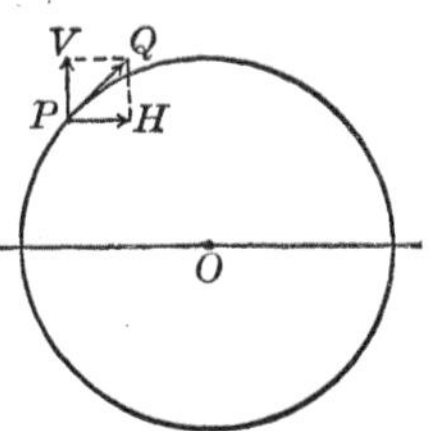

31. In Ex. 30 find a formula for the horizontal speed of P at any instant; that is, when P is at any point on the rim.

32. If two tangents are drawn from $(a, 0)$ to the circle $x^2 + y^2 = r^2$, find the equation of the circle inscribed in the triangle formed by the two tangents and the chord of contact.

33. A ray of light parallel to XO enters a circular disk of glass of radius 13 cm. at $P\,(12, 5)$ and is refracted in the direction PQ. If the direction PQ is determined by the law $\sin\theta = \frac{125}{91}\sin\phi$, PN being the normal at P, find the equation of PQ.

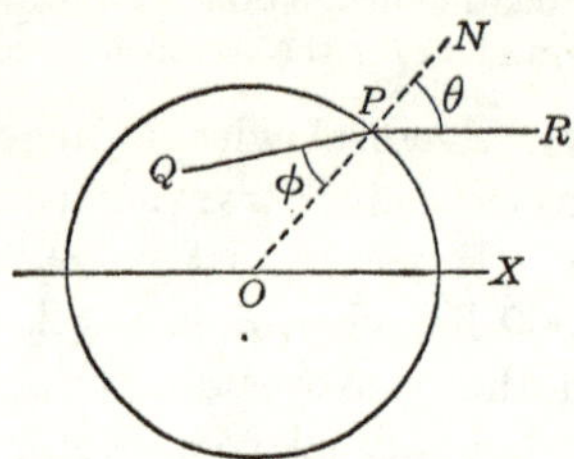

34. Show that the three circles $x^2 + y^2 - 6x - 12y + 41 = 0$, $x^2 + y^2 - 10x - 8y = -37$, and $x^2 + y^2 - 14x - 4y = -49$ have two points in common.

35. Find the equation of a circle passing through the point $(1, 3)$ so that the radical axis of this circle and the circle $x^2 + y^2 - 8x + 7y = 10$ is $2x - 3y = 6$.

36. Find the locus of a point which moves so that the square of its distance from a given point is equal to its distance from a given line.

37. A point moves so that the sum of the squares of its distances from the sides of an equilateral triangle is constant. Show that the locus of the point is a circle.

38. Given an equilateral triangle ABC, find the locus of the point P which moves so that $\overline{PA}^2 = \overline{PB}^2 + \overline{PC}^2$.

39. If the point P moves so that the tangents from P to two given circles have a constant ratio, the locus of P is a circle passing through the common points of the given circles.

40. In the triangle ABC it is given that $AB = 10$ in., that $AC = 12$ in., and that AC revolves about A while AB remains fixed in position. Find the locus of the mid point of BC.

41. Given any two nonconcentric circles $S = 0$ and $S' = 0$, and k any constant, then the center of the circle $S + kS' = 0$ is on the line of centers of $S = 0$ and $S' = 0$.

42. In Ex. 41 show that the line of centers is divided by the center of $S + kS' = 0$ in the ratio $k : 1$.

What interpretation is to be given to the problem if $k = 0$? if $k = 1$?

43. The equation of the circle determined by the three given points (x_1, y_1), (x_2, y_2), and (x_3, y_3) may be written in determinant form as follows:

$$\begin{vmatrix} x^2 + y^2 & x & y & 1 \\ x_1^2 + y_1^2 & x_1 & y_1 & 1 \\ x_2^2 + y_2^2 & x_2 & y_2 & 1 \\ x_3^2 + y_3^2 & x_3 & y_3 & 1 \end{vmatrix} = 0.$$

Comparing the coefficients of x^2 and y^2 in the expansion of the determinant, show by § 92 that the equation represents a circle. Then show that the equation is satisfied when x_1 is substituted for x, and y_1 for y, and similarly for the coordinates of (x_2, y_2) and (x_3, y_3). The exercise may be omitted by those who are not familiar with determinants.

44. Determine the point from which the tangents to the circles $x^2 + y^2 - 2x - 6y + 6 = 0$ and $x^2 + y^2 - 2y - 2x + 14 = 0$ are each equal to $\sqrt{102}$.

45. Find the equation of the locus of the point P which moves so that the tangent from P to the circle $x^2 + y^2 - 6x + y = 7$ is equal to the distance from P to the point $(-7, 5)$.

46. The point P moves so that the tangents from P to the circles $x^2 + y^2 - 6x = 0$ and $x^2 + y^2 + 6x - 2y = 6$ are inversely proportional to the radii. Find the locus of P.

47. Find the equation of the locus of the point P which moves so that the distance from P to the point $(6, -1)$ is twice the length of the tangent from P to the circle $x^2 + y^2 - 7x = 17$.

48. Show that the locus of points from which tangents to the circles $x^2 + y^2 - 12x = 0$ and $x^2 + y^2 + 8x - 3y - 4 = 0$ are in the ratio $2 : 3$ is a circle. Find the center of this circle.

49. If three circles have one and the same radical axis, the lengths of the tangents to two of the circles from a point on the third are in a constant ratio.

50. Find the equation of the circle having the center (5, 4) and tangent internally to the circle $x^2 + y^2 - 6x - 8y = 24$.

51. Find the equation of the circle having the center (h, k) and tangent internally to the circle $x^2 + y^2 - 2ax - 2by + c = 0$.

52. If the axes are inclined at an angle ω, the equation of the circle with center (a, b) and radius r is

$$(x - a)^2 + (y - b)^2 + 2(x - a)(y - b)\cos\omega = r^2.$$

Find the equation of the locus of the center of a variable circle satisfying the following conditions:

53. The circle is tangent to two given fixed circles one of which is entirely within the other.

54. The circle is tangent to a line AB and cuts the constant length $2c$ from a line AC perpendicular to AB.

55. The circle is tangent to a given line and passes through a fixed point at a given distance from the given line.

56. Given the two lines $y = mx - 4$ and $y = -\dfrac{1}{m}x + 4$; when m varies, the lines vary and their intersection varies. Find the locus of their intersection.

Regarding the equations as simultaneous, if we eliminate m we obtain the required equation in x and y; for this equation is satisfied by the coordinates of the intersection of the lines, whatever the value of m.

57. Find the equation of the locus of the intersection of the lines $y = mx + \sqrt{m^2 + 2}$ and $y = -\dfrac{1}{m}x + \sqrt{\dfrac{1}{m^2} + 2}$.

To eliminate m transpose the x term in each equation, clear of fractions, square both members, and add.

58. All circles of the system $x^2 + y^2 - 2ax - 2ay + a^2 = 0$ are tangent to both axes.

59. All circles of the system $x^2 + y^2 - 2ax - 4ay + 4a^2 = 0$ are tangent to the line $4y = 3x$ and also to the y axis.

60. Find the equation that represents the system of circles with centers on OX and tangent to the line $y = x$.

CHAPTER VI

TRANSFORMATION OF COORDINATES

104. Change of Axes. The equation of the circle with center $C(2, 3)$ and radius 4 is, as we have seen,

$$(x-2)^2+(y-3)^2=16.$$

Here the y axis and x axis are respectively 2 units and 3 units from C. But if we discard these axes and take a new pair CX, CY through the center C, then the equation of the circle referred to these axes is

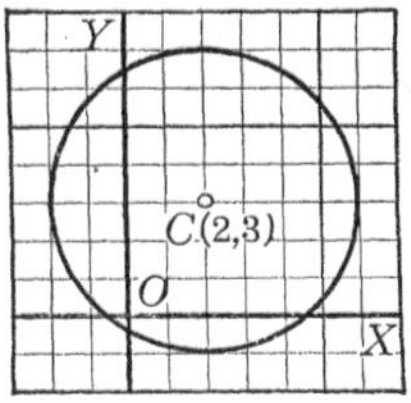

$$x^2+y^2=16,$$

which is simpler than the other.

In general, the coefficients of the equation of a curve depend upon the position of the axes. If the axes are changed, the coefficients are, of course, usually changed also, and it is to be expected that often, when we know the equation of a curve referred to one pair of axes, we may find a new pair of axes for which the equation of the same curve is simpler.

The process of changing the axes of coordinates is called *transformation of coordinates*, and the corresponding change in the equation of the curve is called *transformation of the equation*.

The two transformations explained in this chapter are merely the special cases which are most often needed in our work. The general problem is to change from any pair of axes to any other pair, rectangular or oblique.

Problem. Translation of Axes

105. *To change the origin from the point (0, 0) to the point (h, k), without changing the direction of the axes.*

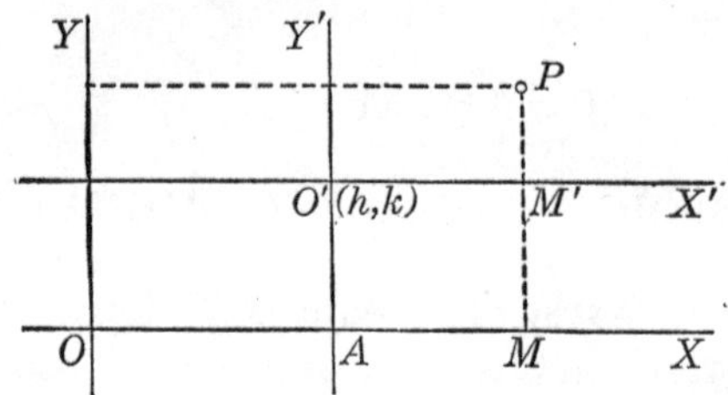

Solution. Let OX, OY be the given axes, O' the point (h, k), $O'X'$, $O'Y'$ the changed axes. Denote by (x, y) the coordinates of any point P referred to the given axes, and by (x', y') the coordinates of P referred to the changed axes.

Then $$x = OA + AM,$$
and $$y = AO' + M'P.$$
That is, $$x = x' + h,$$
and $$y = y' + k.$$

Hence, when we know the equation of a curve referred to one pair of axes, we find its equation referred to another pair of axes through (h, k) parallel to the given axes by putting $x' + h$ for x and $y' + k$ for y in the given equation. No confusion results if in the new equation in x' and y' we drop the primes and write x and y for x' and y'.

That is, *to find the transformed equation write $x + h$ for x and $y + k$ for y in the given equation.*

106. Translation of Axes. The kind of transformation discussed in § 105 is called a *translation.*

It is customary to speak of the *translation of coordinates* or of the *translation of axes.*

107. Simplifying an Equation by Translation. Let us examine the possible effects of translation upon the equation

$$2x^2 - 3xy + y^2 + 5x - 3y - 10 = 0.$$

If the new origin is (h, k), then the new equation is

$$2(x+h)^2 - 3(x+h)(y+k) + (y+k)^2 + 5(x+h) - 3(y+k) - 10 = 0,$$

or $$2x^2 - 3xy + y^2 + (4h - 3k + 5)x + (-3h + 2k - 3)y + 2h^2 - 3hk + k^2 + 5h - 3k - 10 = 0.$$

Only the x term, the y term, and the constant term depend on h and k. To make the equation simpler, let us determine h and k so that the coefficients of two of these terms, say the x term and y term, vanish; that is, so that $4h - 3k + 5 = 0$ and $-3h + 2k - 3 = 0$.

This gives $h = 1$, $k = 3$; and the new equation, referred to axes through (1, 3) and parallel to the old axes, is

$$2x^2 - 3xy + y^2 = 12.$$

108. Equation $Ax^2 + By^2 + Cx + Dy + E = 0$. We shall frequently meet an equation of the second degree which has an x^2 term and a y^2 term but no xy term.

The following example shows an easier method than the one given above for transforming such an equation into an equation which has no terms of the first degree.

For example, consider the equation $2x^2 - 6y^2 - 16x - 12y = 1$. We may first write it in the form

$$2(x^2 - 8x) - 6(y^2 + 2y) = 1,$$

and then $$2(x-4)^2 - 6(y+1)^2 = 1 + 32 - 6 = 27.$$

It is now obvious that the desired new origin is $(4, -1)$; for, by writing $x + 4$ for x and $y - 1$ for y, we have

$$2x^2 - 6y^2 = 27.$$

PROBLEM. ROTATION OF AXES

109. *To change the direction of the axes without changing the origin, both sets of axes being rectangular.*

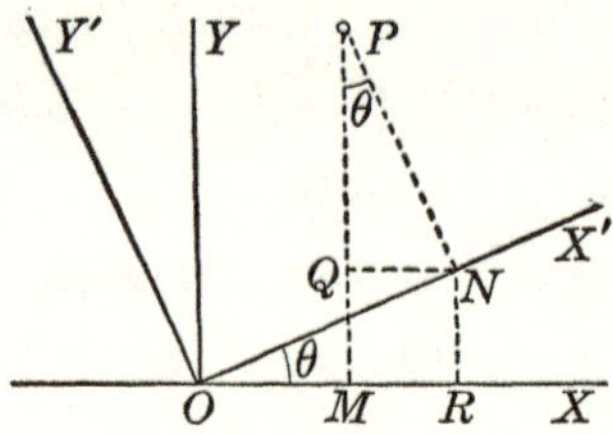

Solution. Let (x, y) be a point P referred to the given rectangular axes OX, OY, and let (x', y') be the same point referred to the changed axes OX', OY'. Then we have

$$OM = x, \quad MP = y, \quad ON = x', \quad NP = y'.$$

Let θ represent the angle XOX'. Draw NQ perpendicular to PM, and NR perpendicular to OX. Then angle $QPN = \theta$, since their sides are perpendicular each to each.

Hence $\qquad OM = OR - QN = ON \cos\theta - NP \sin\theta,$

that is, $\qquad \boldsymbol{x = x' \cos\theta - y' \sin\theta.}$

Also $\qquad MP = RN + QP = ON \sin\theta + NP \cos\theta,$

that is, $\qquad \boldsymbol{y = x' \sin\theta + y' \cos\theta.}$

Hence, when we know the equation of a curve referred to one pair of rectangular axes, we can find its equation referred to another pair of rectangular axes with the same origin and making the angle θ with the given axes.

To find the transformed equation write $x \cos\theta - y \sin\theta$ for x and $x \sin\theta + y \cos\theta$ for y in the given equation.

110. Rotation of Axes. The kind of transformation discussed in § 109 is called a *rotation*.

THEOREM. ELIMINATION OF THE xy TERM

111. *In rectangular coordinates it is always possible to determine a rotation which transforms any equation of the second degree into an equation which lacks the xy term.*

Proof. Let $ax^2 + 2hxy + by^2 + cx + dy + e = 0$ represent any equation of the second degree. Rotation of the axes through an angle θ transforms this equation into

$$a(x\cos\theta - y\sin\theta)^2 + 2h(x\cos\theta - y\sin\theta)(x\sin\theta + y\cos\theta)$$
$$+ b(x\sin\theta + y\cos\theta)^2 + (\text{terms of lower degree}) = 0.$$

In this equation, collecting all the xy terms, we have as the coefficient of xy

$$-2a\sin\theta\cos\theta + 2h\cos^2\theta - 2h\sin^2\theta + 2b\sin\theta\cos\theta.$$

Then, since by trigonometry $2\sin\theta\cos\theta = \sin 2\theta$ and $\cos^2\theta - \sin^2\theta = \cos 2\theta$ (page 286), the coefficient of xy is

$$-(a-b)\sin 2\theta + 2h\cos 2\theta.$$

Then, as is always possible, we choose θ so that

$$-(a-b)\sin 2\theta + 2h\cos 2\theta = 0;$$

that is, so that $$\tan 2\theta = \frac{2h}{a-b}.$$

The transformed equation will then have no xy term.

For example, to transform the equation $3x^2 + 3xy - y^2 = 1$ into an equation which has no xy term, we have $\tan 2\theta = \frac{3}{4}$. Since we must use $\sin\theta$ and $\cos\theta$, we substitute $\frac{3}{4}$ for $\tan 2\theta$ in the formula $\tan 2\theta = \dfrac{2\tan\theta}{1-\tan^2\theta}$ (page 286). Clearing of fractions and simplifying, we have $3\tan^2\theta + 8\tan\theta - 3 = 0$, whence $\tan\theta = \frac{1}{3}$ or -3. Choosing $\tan\theta = \frac{1}{3}$, we find by calculation that $\sin\theta = 1/\sqrt{10}$ and $\cos\theta = 3/\sqrt{10}$. Completing the transformation of the equation $3x^2 + 3xy - y^2 = 1$ by writing $(3x - y)/\sqrt{10}$ for x and $(x + 3y)/\sqrt{10}$ for y and simplifying, we have the transformed equation $7x^2 - 3y^2 - 2 = 0$.

Exercise 33. Transformation of Equations

Transform each of the following equations by means of a translation, making O′ the new origin:

1. $x^2 - 2y^2 - 6x + 8y = 9$; $O'(3, 2)$.
2. $y^2 - 4x + 4y + 8 = 0$; $O'(1, -2)$.
3. $(x - a)^2 + (y + a)^2 = xy$; $O'(a, -a)$.
4. $x^2 - 6xy + y^2 - 6x + 2y + 1 = 0$; $O'(0, -1)$.
5. $3x^2 - 2xy - y^2 + 14x + 6y = 6$; $O'(-1, 4)$.

Transform each of the following equations into an equation which has no first-degree terms:

6. $x^2 + y^2 - 6x + 2y = 6$.
7. $2x^2 - 2y^2 - 8x - 4y = 5$.
8. $3x^2 + xy - 2y^2 + 4x + 9y = 15$.
9. $x^2 - 3xy - 3y^2 + 6x + 12y = 16$.

Transform each of the following equations by means of a rotation of the axes through the acute angle θ:

10. $x^2 - y^2 = 4$; $\theta = 45°$.
11. $(x + y - 2)^2 = 4xy$; $\theta = 45°$.
12. $2x^2 + 24xy - 5y^2 = 8$; $\theta = \sin^{-1}\frac{3}{5}$.
13. $x^2 - 2xy + 2y^2 = 1$; $\theta = \tan^{-1} 2$.

Transform each of the following equations into an equation which has no xy term:

14. $xy = 12$.
15. $x^2 + y^2 + xy = 3$.
16. $x^2 - 2xy + y^2 = \sqrt{2}(x + y)$.
17. $3x^2 + 12xy + 8y^2 = 5$.

18. Show that there is no translation of axes which transforms the equation $y^2 - 6y - 2x + 11 = 0$ into an equation with no first-degree terms. Find a translation which removes two of the terms of the given equation.

CHAPTER VII

THE PARABOLA

112. Conic Section. The locus of a point which moves so that the ratio of its distances from a fixed point and a fixed line is constant is called a *conic section*, or simply a *conic*.

We shall designate the moving point by P, the fixed line by l, the fixed point by F, the distance of P from F by d, the distance of P from l by d', and the constant ratio $d : d'$ by e. The fixed line l is called the *directrix* of the conic, the fixed point F is called the *focus* of the conic, and the constant ratio e is called the *eccentricity* of the conic.

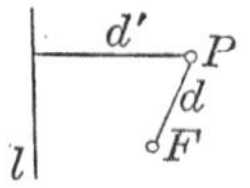

Conic sections are so called because they were first studied by synthetic geometry as plane sections of a right circular cone. We shall now study conic sections by means of coordinates, a plan much simpler than the ancient geometric plan.

The proof that conic sections, as defined above, are plane sections of a right circular cone is not given in this book.

113. Classes of Conics. The form of a conic depends on the value of the ratio $d : d'$.

If $e = 1$, that is, if $d = d'$, the conic is called a *parabola*.

If $e < 1$, the conic is called an *ellipse*.

If $e > 1$, the conic is called a *hyperbola*.

As we shall see later, the circle is a special case of the ellipse.

114. Parabola. The locus of a point P which moves so that its distances from a fixed point F and a fixed line l are equal is called a *parabola*.

PROBLEM. EQUATION OF THE PARABOLA

115. *To find the equation of the parabola when the line through the focus perpendicular to the directrix is the x axis and the origin is midway between the focus and the directrix.*

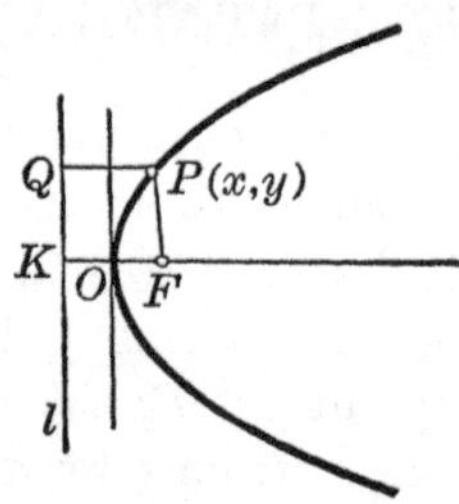

Solution. By the definition in § 114 it follows that a locus is a parabola if it satisfies the condition $FP = QP$.

Denote by $2p$ the fixed distance KF from l to F. Then O, the mid point of KF, being equidistant from l and F, is a point on the parabola. Taking the origin at the point O and the x axis along KF, the fixed point F is $(p, 0)$; and if $P(x, y)$ is any point on the parabola, the equation of the parabola is found from the condition

$$FP = QP;$$

that is, $$\sqrt{(x-p)^2 + y^2} = p + x.$$

Therefore $$\mathbf{y^2 = 4px.}$$

This equation of the parabola is the basis of our study of the properties of this curve.

116. Position of the Focus and Directrix. The focus of the parabola $y^2 = 4px$ is evidently the point $(p, 0)$, and the directrix is the line $x = -p$.

Thus $(6, 0)$ is the focus of the parabola $y^2 = 24x$, and the directrix is the line $x = -6$.

117. Drawing the Parabola. Since the equation of the parabola $y^2 = 4px$ may be written $y = \pm 2\sqrt{px}$, the parabola is evidently symmetric with respect to OX. From the equation it is also apparent that as x increases, y increases numerically.

If p is positive, the curve lies on the positive side of OY because any negative value of x makes y imaginary.

If p is negative, F lies on the left of l and the curve lies on the left of OY.

To draw a parabola from its equation, plot a few of its points and connect these points by a smooth curve.

118. Axis and Vertex. The line through F perpendicular to the directrix is called the *axis* of the parabola, and the point O where the parabola cuts its axis is called the *vertex*.

119. Focus on the y Axis. When the y axis is taken as the axis of the parabola, the origin being at the vertex as in this figure, by analogy with the result of § 115 the equation of the parabola is

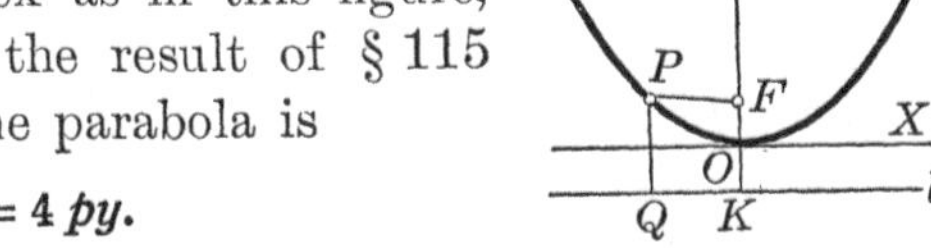

$$x^2 = 4py.$$

120. Chord. A line joining any two points on a conic is called a *chord* of the conic. A chord passing through the focus of a conic is called a *focal chord*. A line from the focus of a conic to a point on the curve is called a *focal radius*.

The focal chord perpendicular to the axis is called the *focal width* or *latus rectum*.

For example, AD, BC, and BE are chords of this parabola; AD and BC are focal chords; FB is a focal radius; and AD is the focal width.

Exercise 34. Equations of Parabolas

Draw each of the following parabolas; find the focal width, the coordinates of the focus, and the equation of the directrix; mark the focus and draw the directrix in each figure:

1. $y^2=8x$.	**3.** $y^2=12x$.	**5.** $6x=-y^2$.	**7.** $x^2=-8y$.
2. $y^2=-8x$.	**4.** $x=2y^2$.	**6.** $x^2=8y$.	**8.** $2x^2=9y$.

In the equation $y^2=4px$ find the value of p under each of the following conditions:

9. The parabola passes through the point (2, 6).

10. The focal width is 16.

11. The distance from the vertex to the focus is 3.

12. The focus is the point (– 5, 0).

Under each of the following conditions find the equation of the parabola whose vertex is at the origin and whose axis is one of the coordinate axes:

13. The line from (– 2, 5) to (– 2, – 5) is a chord.

14. The line from (– 2, 6) to (2, 6) is a chord.

15. The focus is (0, – 3).

16. The focus is on the line $3x+4y=12$.

17. The directrix is the line $y=6$.

18. Show that the focal width of the parabola $y^2=4px$ is $4p$.

19. For what point on the parabola $x^2=-12y$ is the ordinate twice the abscissa?

20. From the definition of the parabola give a geometric construction by which points on the parabola may be found when the directrix and focus are given.

21. The distance from the focus to any point $P_1(x_1, y_1)$ on the parabola $y^2=4px$ is $p+x_1$.

22. If the distance from the focus to a point P on the parabola $y^2=4x$ is 10, find the coordinates of P.

23. In the parabola $y^2 = 4px$ an equilateral triangle is inscribed so that one vertex is at the origin. Find the length of one side of the triangle.

A figure is inscribed in a curve if all the vertices lie on the curve.

24. Given the parabola $y^2 = 16x$, find the equation of the circle with center (12, 0) and passing through the ends of the focal width. By finding the points of intersection of the circle and parabola, show that the circle and parabola are tangent.

Draw the graphs of each of the following pairs of equations and find the common points in each case:

25. $y^2 = 9x$; $3x - 7y + 30 = 0$.

26. $y^2 = 3x$; $x - 4y + 12 = 0$.

27. $x^2 = 18y$; $2x - 6y + 3 = 0$.

28. $y^2 = 12x$; $x - y = 9$.

29. $x^2 + y^2 - 4x = 4$; $x = y^2$.

30. $x^2 = y$; $y^2 = x$.

31. $x^2 = 4$; $y^2 = 9$.

32. $x^2 = 4y$; $y^2 = 9x$.

33. $x^2 = 9y$; $y^2 = 2x$.

34. $x^2 = y$; $y = 1$.

35. Draw the parabola $y^2 = 6x$ and the chords made by the parallel lines $y = x$, $y = x - \frac{8}{3}$, and $y = x - \frac{9}{2}$. Show that the mid points of these chords lie on a straight line.

36. Draw the parabola $x^2 = -4y$ and on it locate the point P whose abscissa is 6. Join P to the focus F and prove that the circle whose diameter is FP is tangent to the x axis.

From the definition of the parabola find the equation of the parabola in each of the following cases:

37. The y axis lies along the directrix and the x axis passes through the focus.

38. The x axis is the axis of the parabola and the origin is the focus.

39. The directrix is the line $x = 6$ and the focus is the point (4, 2).

40. The directrix is the line $3x = 4y - 3$ and the focus is the point (1, 1).

Problem. More General Equation of the Parabola

121. *To find the equation of the parabola with its axis parallel to the x axis and with its vertex at the point* (h, k).

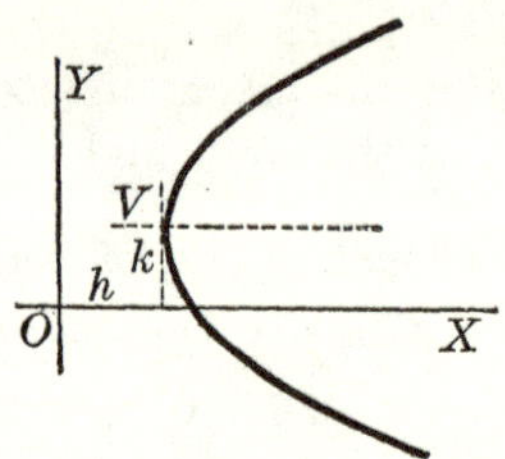

Solution. Letting $V(h, k)$ be the vertex, we are to find the equation of the parabola referred to the axes OX, OY.

If we suppose for a moment that the axes are taken through V parallel to OX and OY, the equation of the parabola referred to these axes is $y^2 = 4px$. Referred to V as origin, O is the point $(-h, -k)$. If we move the origin to O (§ 105), we must write $x - h$ for x and $y - k$ for y in the equation $y^2 = 4px$, and the equation becomes

$$(y - k)^2 = 4p(x - h).$$

This is therefore the equation of the parabola referred to the axes OX, OY.

122. Corollary. *The equation* $Ay^2 + By + Cx + D = 0$, *where* $A \neq 0$ *and* $C \neq 0$, *represents a parabola with its axis parallel to the x axis.*

For, by completing the square with respect to the terms $Ay^2 + By$, this equation may be written in the form $(y - k)^2 = 4p(x - h)$.

For example, the equation $3y^2 - 12y - 5x + 2 = 0$ may be written $3(y^2 - 4y) = 5x - 2$, or $3(y - 2)^2 = 5x + 10$, or $(y - 2)^2 = \frac{5}{3}(x + 2)$.

This is therefore the equation of a parabola with its axis parallel to OX and with its vertex at the point $(-2, 2)$.

123. Axis Parallel to the y Axis. When the axis of a parabola is the y axis and the vertex is the origin, the equation of the parabola (§ 119) is $x^2 = 4py$. Therefore by § 121 the equation of a parabola with its axis parallel to OY and with the vertex (h, k) is

$$(x - h)^2 = 4p(y - k);$$

and if $A \neq 0$ and $C \neq 0$, every equation of the form

$$Ax^2 + Bx + Cy + D = 0 \qquad \S\ 122$$

represents a parabola with its axis parallel to the y axis.

Exercise 35. Axis Parallel to *OX* or to *OY*

Find the vertex V of each of the following parabolas, draw new axes through V parallel to the given axes, find the new equation, and draw the parabola:

1. $y^2 - 4y - 6x + 10 = 0$.
2. $2y^2 + 12y + 3x + 3 = 0$.
3. $3y^2 + 12y + 16 = 4x$.
4. $y^2 = 3x + 2y + 5$.
5. $y = 2x^2 - 6x + 3$.
6. $5x^2 - 5x = 4y$.

7. What are the coordinates of the focus of the parabola $2y^2 = 8y + 3x + 1$?

8. Find the focal width of the parabola $3x^2 - 6x = 4y - 11$.

9. Show that the two parabolas $x^2 - 2x = 5y - 11$ and $y^2 = 4y + 5x - 9$ have the same vertex, and find the other point of intersection.

10. The equations $y = ax^2 + bx + c$, $x = py^2 + qy + r$ represent parabolas with axes parallel to OY and OX respectively.

Find the equation of the parabola, given that:

11. The vertex is $(2, -3)$ and the focus is $(6, -3)$.

12. The focus is $(6, 8)$ and the directrix is the line $y = -2$.

13. The vertex is $(4, 3)$ and the directrix is the line $x = 6$

14. The focus is $(0, 0)$ and the vertex is $(p, 0)$.

124. Tangent. We have hitherto regarded a tangent to a circle as a line which touches the circle in one point and only one. This definition of tangent does not, however, apply to curves in general.

Thus, we may say that l is tangent to this curve at A, although l cuts the curve again at B.

If c is any curve and P any point on it, we define the tangent to c at P as follows:

Take another point Q upon the curve, and draw the secant PQ. Letting Q move along the curve toward P, as Q approaches P the secant PQ turns about P and approaches a definite limiting position PT. The line PT is then said to be tangent to the curve c at P.

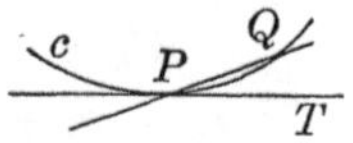

The secant PQ cuts the curve in the two points P and Q. As the secant approaches the position of the tangent PT, the point Q approaches the point P; and when the secant coincides with the tangent, Q coincides with P.

We therefore say that *a tangent to a curve cuts the curve in two coincident points* at the point of tangency.

The tangent may even cross the curve at P, as shown in this figure, but in such a case the tangent cuts the curve in three coincident points, as is evident if the secant PQ is extended to cut the curve again at the left of the point P.

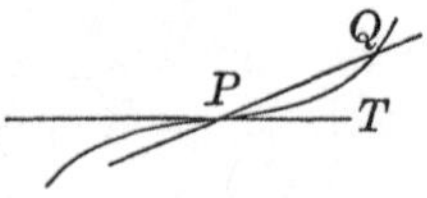

125. Slope of a Curve. The slope of the tangent at a point P on a curve is called also the *slope of the curve at the point P.*

Thus, in the case of the parabola shown on the opposite page, the slope of the curve at the point P_1 is the slope of the tangent P_1T.

While the slope of a straight line does not vary, the slope of a curve is different at different points of the curve.

PROBLEM. SLOPE OF THE PARABOLA

126. *To find the slope of the parabola* $y^2 = 4px$ *at any point* $P_1(x_1, y_1)$ *on the parabola.*

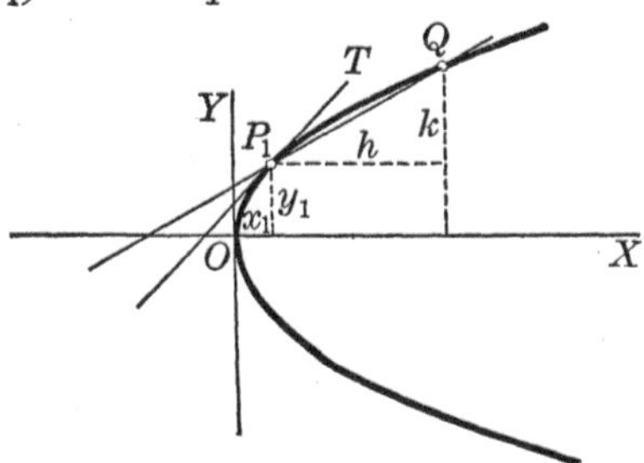

Solution. Take another point $Q(x_1 + h, y_1 + k)$ on the parabola. Then the slope of the secant P_1Q is k/h, and the slope m of the tangent at P_1 is the limit of the slope of the secant as Q approaches P_1. That is,

$$m = \lim_{Q \to P_1} \frac{k}{h}.$$

Since P_1 and Q are both on the parabola, we have

$$y_1^2 = 4px_1 \tag{1}$$

and

$$(y_1 + k)^2 = 4p(x_1 + h). \tag{2}$$

Subtracting (1) from (2), we have

$$k(2y_1 + k) = 4ph;$$

whence

$$k/h = 4p/(2y_1 + k).$$

Now when $Q \rightarrow P_1$, $k \rightarrow 0$; and hence $\lim k/h = 4p/2y_1$. But $\lim k/h = m$, the slope of the tangent P_1T.

Hence

$$m = \frac{2p}{y_1}.$$

That is, *the slope of the parabola* $y^2 = 4px$ *at any point is equal to* $2p$ *divided by the ordinate of the point.*

PROBLEM. TANGENT TO THE PARABOLA

127. *To find the equation of the tangent to the parabola* $y^2 = 4px$ *at any point* $P_1(x_1, y_1)$ *on the parabola.*

Solution. Since the tangent passes through the point (x_1, y_1) and has the slope $2p/y_1$, its equation is

$$y - y_1 = \frac{2p}{y_1}(x - x_1). \qquad \S\ 69$$

Clearing of fractions, we have

$$y_1 y - y_1^2 = 2px - 2px_1;$$

and since (x_1, y_1) is on the parabola $y^2 = 4px$, we have $y_1^2 = 4px_1$, and the equation takes the convenient form

$$\boldsymbol{y_1 y = 2p(x + x_1).}$$

For example, the equation of the tangent to the parabola $y^2 = 12x$ at the point $(\frac{1}{3}, -2)$ is $-2y = 6(x + \frac{1}{3})$, or $3x + y + 1 = 0$.

128. Slope of any Curve. The method of finding the slope of the parabola (§ 126) may be used to find the slope of any curve when the equation of the curve is known, thus solving a problem not only of great historic interest but of the greatest practical importance.

Finding the limit of k/h, the slope of the secant, turns out to be a difficult matter in the case of many curves, and the treatment of such cases is explained in the calculus.

129. Subtangent and Subnormal. The projections upon the x axis of the segments of the tangent and the normal included between the point of tangency P and the x axis are called the *subtangent* and the *subnormal* for the point P.

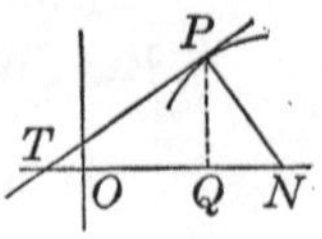

Thus, in the figure, if PT is the tangent at P and PN is the normal, TQ is the subtangent, and QN the subnormal.

Exercise 36. Tangent at the Point (x_1, y_1)

In each of the following parabolas show that P is on the curve, and find the equations of the tangent and the normal at P:

1. $y^2 = 8x$; $P(2, 4)$.
2. $y^2 = -10x$; $P(-3.6, -6)$.
3. $y^2 + x = 0$; $P(-1, 1)$.
4. $x = 9y^2$; $P(9, -1)$.

5. Find the intercepts of the tangent to the parabola $y^2 = 12x$ at the point $(\frac{16}{3}, 8)$.

6. Find the intercepts of the tangent to the parabola $y^2 = 4px$ at the point $P_1(x_1, y_1)$. Show that the x intercept is $-x_1$ and that the y intercept is $\frac{1}{2}y_1$. Draw the figure.

7. From Ex. 6 find a geometric construction for the tangent to any parabola at any point on the parabola.

8. Find the equations of the tangents to the parabola $y^2 = -16x$ at the ends of the focal width. Show that these tangents are perpendicular to each other and that they meet on the directrix.

9. The tangents to any parabola at the ends of the focal width meet at right angles on the directrix.

10. A tangent is drawn to the parabola $y^2 = 4x$ at the point $(9, 6)$. Find the subtangent and the subnormal.

11. A tangent is drawn to the parabola $y^2 = 4px$ at the point (x_1, y_1). Find the subtangent and the subnormal.

12. On the parabola $y^2 = 10x$ find the point (a, b) for which the subtangent is 12.

13. On the parabola $3x + 8y^2 = 0$ find the point P such that the tangent at P has equal intercepts.

14. On the parabola $2y^2 = 9x$ find the point P such that the tangent at P passes through the point $(-6, 3)$.

15. Suppose that a ray of light from the focus strikes the parabola $y^2 = 4x$ at the point $(9, 6)$. Draw the figure, draw the reflected ray, and find the equation of this ray.

130. Properties of the Parabola. One important property of the parabola was proved in § 126; namely, that the slope of the curve $y^2 = 4px$ at the point (x_1, y_1) is $2p/y_1$. Other properties were expressed in Exercises 34–36, and the parabola has, of course, a great many more. The following properties, arising in connection with the tangent to the parabola $y^2 = 4px$ at the point (x_1, y_1), are of special importance and should be carefully studied:

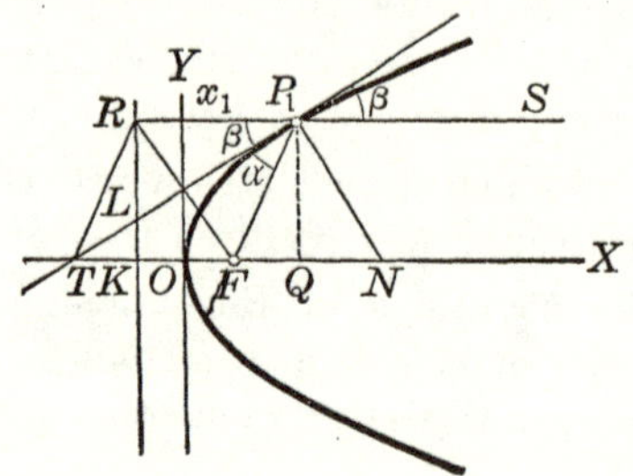

1. *The subtangent TQ is bisected at the origin.*

Proof. The equation of the tangent at $P_1(x_1, y_1)$ is $y_1 y = 2p(x + x_1)$. The x intercept OT is found by letting $y = 0$ in this equation and finding x. In this way we find that $x = -x_1 = OT$. But $OQ = x_1$, and hence OT and OQ have the same length, so that O is the mid point of TQ.

2. *Taking F as the focus and P_1R as the perpendicular to the directrix KR, the tangent at P_1 bisects the angle RP_1F.*

Proof. We have $P_1R = P_1F$. § 114

Furthermore, $RP_1 = p + x_1$,

and, by property 1, $TF = TO + OF = OQ + OF$

$= x_1 + p$.

Therefore $RP_1 = TF = FP_1$.

Hence TFP_1R is a rhombus and $\beta = \alpha$.

3. *If RP_1 is produced to any point S, FP_1 and P_1S make equal angles with the normal P_1N.*

Proof. We have $\alpha = \beta$ from property 2. But it is evident from the figure that angle FP_1N is the complement of α, and that angle NP_1S is the complement of β.

Hence angle $FP_1N =$ angle NP_1S.

This is the important *reflection property* of the parabola.

Rays of light from F are reflected parallel to the axis. This fact is used in making headlights, searchlights, and reflecting telescopes. Sound or heat from F is also reflected parallel to the axis.

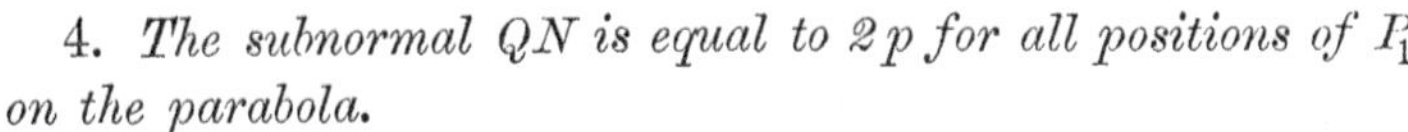

4. *The subnormal QN is equal to $2p$ for all positions of P_1 on the parabola.*

Proof. Since TFP_1R is a rhombus, RF is perpendicular to TP_1. Hence RF is parallel to P_1N.

Therefore the triangles RKF and P_1QN are congruent, and

$$QN = KF = 2p.$$

5. *Through any point (h, k) in the plane there are two tangents to the parabola $y^2 = 4px$.*

Proof. The tangent at (x_1, y_1) is $y_1 y = 2p(x + x_1)$, and this passes through (h, k) if and only if $y_1 k = 2p(h + x_1)$.

Since $y_1^2 = 4px_1$, the second equation reduces to

$$y_1 k = 2p\left(h + \frac{y_1^2}{4p}\right),$$

or

$$y_1^2 - 2ky_1 + 4ph = 0,$$

which is satisfied by two values of y_1. Hence there are two points (x_1, y_1) the tangents at which pass through (h, k).

The two tangents may be distinct or coincident, real or imaginary.

Exercise 37. Properties of the Parabola

Draw the figure given on page 126, and prove that:

1. The tangent at P_1 bisects RF at right angles.

2. The lines TP_1, RF, and the y axis meet in a point.

3. The line joining F and L is perpendicular to FP_1.

4. The focus is equidistant from T, P_1, and N.

5. The focus is equidistant from the points in which the tangent cuts the directrix and the focal width produced.

6. The segment RF is the mean proportional between the segment FP_1 and the focal width.

The student will probably have difficulty in finding a geometric proof of this statement, but the analytic proof is simple.

7. $\overline{TK}^2 \cdot KO = TF \cdot d^2$, where d is the distance from K to TP_1.

8. The proof of property 2, § 130, is geometric. Give an analytic proof, using the slopes of RP_1, TP_1, FP_1.

9. Give an analytic proof of property 4, § 130.

As the first step the student may find the x intercept of the normal.

10. A line revolving about a fixed point on the axis of the parabola $y^2 = 4px$ cuts the parabola in P and Q. Show that the product of the ordinates of P and Q is constant, and likewise that the product of the abscissas is constant.

11. The line l revolves about the origin and cuts the parabolas $y^2 = 4px$ and $y^2 = 4qx$ again in R and S. Show that the ratio $OR : OS$ is constant for all positions of l.

12. Using property 1, § 130, find a geometric construction for the tangent to a parabola at a point P_1 on the curve.

13. Consider Ex. 12, using property 4, § 130.

14. Find a geometric construction for the tangent to a parabola from a point not on the curve, using Ex. 1 above.

15. There are three normals from (h, k) to the parabola $y^2 = 4px$, and either one or three of them are real.

PROBLEM. TANGENT WITH A GIVEN SLOPE

131. *To find the equation of the line which has the slope m and is tangent to the parabola $y^2 = 4px$.*

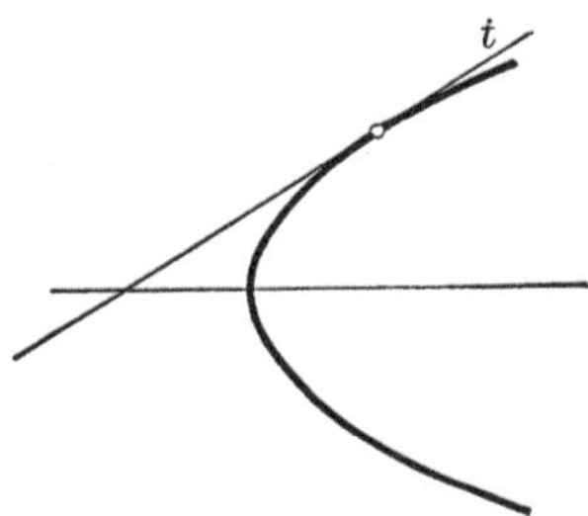

Solution. Let $y = mx + k$ represent the tangent t whose slope is m. The problem then reduces to finding k from the condition that t is tangent to the parabola; that is, that t cuts the parabola in two coincident points (§ 124).

To find the common points of t and the curve we solve the equations $y = mx + k$ and $y^2 = 4px$ as simultaneous.

Then $$(mx + k)^2 = 4px;$$

that is, $$m^2x^2 + 2(km - 2p)x + k^2 = 0.$$

The roots of this quadratic in x are the abscissas of the common points of t and the parabola. But since these points coincide, the roots are equal. Now the condition under which any quadratic $Ax^2 + Bx + C = 0$ has equal roots is that $B^2 - 4AC = 0$. Therefore we have

$$4(km - 2p)^2 - 4k^2m^2 = 0;$$

whence $k = p/m$, and the equation of the tangent is

$$y = mx + \frac{p}{m}.$$

This is called the *slope equation* of the tangent to the parabola.

Exercise 38. Tangents with Given Slopes

Find the equations of the tangents to the following parabolas under the specified conditions:

1. $y^2 = 16x$, the tangent having the slope 2.

2. $y^2 = 12x$, the tangent having the slope $-\frac{1}{3}$.

3. $y^2 = 4x$, the tangent being parallel to $2x + 3y = 1$.

4. $3y^2 = x$, the tangent being perpendicular to $x + y = 0$.

5. $3y^2 + 16x = 0$, two perpendicular tangents, one of the two points of contact being $(-3, 4)$.

6. $y^2 = 12x$, through the point $(-2, -5)$.

The given point is not on the parabola. Let $y = mx + 3/m$ be the tangent, and find m from the condition that $(-2, -5)$ is on the tangent.

7. $x = 2y^2$, from the point $(8, -2)$.

8. $2y^2 + 7x = 0$, from the point $(-3, \frac{17}{8})$.

9. $3y^2 = -4x$, having equal intercepts.

10. Show that the line $y = -3x + 2$ is tangent to the parabola $y^2 = -24x$, and find the point of contact.

11. Find the point of contact of the line which is tangent to the parabola $y^2 = 3x$ and has the slope $\frac{1}{2}$.

12. Find in terms of m the coordinates of the point of contact of the tangent with slope m to the parabola $y^2 = 4px$.

13. Show that the intercepts of the tangent with slope m to the parabola $y^2 = 4px$ are $-p/m^2$ and p/m.

14. A line with slope m is tangent to the parabola $5y^2 = 8x$ and forms with the axes a triangle of area 10. Find m.

15. A line with slope m is tangent to the parabola $3y^2 = 80x$ and is also tangent to the circle $x^2 + y^2 = 9$. Find m.

16. If the line $x - 2y + 12 = 0$ is tangent to the parabola $y^2 = cx$, find c.

17. If two tangents to a parabola are perpendicular to each other, they meet on the directrix.

PROBLEM. PATH OF A PROJECTILE

132. *To find the path of a projectile when the initial velocity is given, disregarding the resistance of the air.*

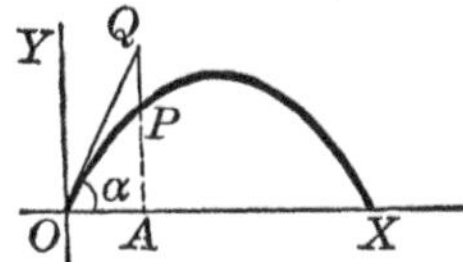

Solution. Take as axes the horizontal and vertical lines through the initial position of the projectile. Let the initial velocity be v, directed at the angle α with OX.

Without the action of gravity the projectile would continue in the straight line OQ, and after t seconds would reach a point Q distant vt from O; that is, $OQ = vt$.

But the action of gravity deflects the projectile from OQ by a vertical distance PQ, which in t seconds is $\frac{1}{2}gt^2$.

It is assumed that the student is familiar with the fact that a body falls from rest $\frac{1}{2}gt^2$ feet in t seconds; and that a body projected upwards loses in t seconds $\frac{1}{2}gt^2$ feet from the height it would have attained without the action of gravity.

Therefore the coordinates OA and AP of the position P of the projectile after t seconds are

$$x = vt\cos\alpha,$$

and

$$y = vt\sin\alpha - \tfrac{1}{2}gt^2.$$

That is, eliminating t, we have

$$y = x\tan\alpha - \frac{g}{2v^2\cos^2\alpha}x^2.$$

But this equation is of the form $Ax^2 + Bx + Cy + D = 0$, which is the equation of a parabola (§ 123).

Therefore the path of a projectile is a parabola.

PROBLEM. MID POINTS OF PARALLEL CHORDS

133. *To find the locus of the mid points of a system of parallel chords of a parabola.*

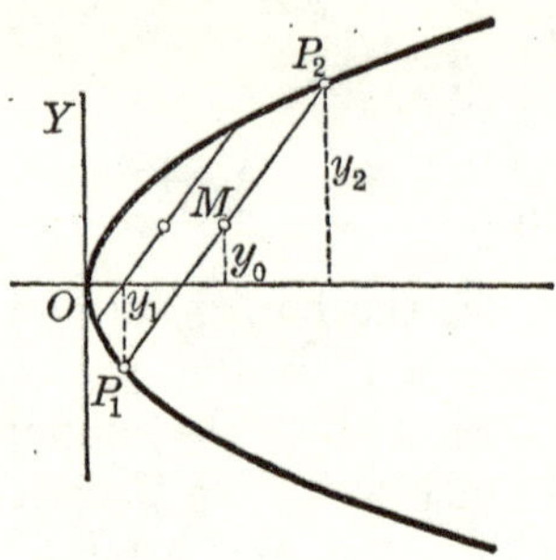

Solution. Represent the parabola by $y^2 = 4px$, the slope of the parallel chords by m, and any one of the parallel chords, say P_1P_2, by $y = mx + b$.

To find the coordinates of P_1 and P_2 we solve the equations $y^2 = 4px$ and $y = mx + b$ as simultaneous. Substituting $y^2/4p$ for x in the equation $y = mx + b$, we have

$$my^2 - 4py + 4pb = 0.$$

The roots of this quadratic are the ordinates y_1 and y_2 of P_1 and P_2; and the ordinate y_0 of the mid point M of P_1P_2 is half their sum (§ 27); that is, $y_0 = \frac{1}{2}(y_1 + y_2)$.

But the sum of the roots of the quadratic is $4p/m$.

Therefore $\qquad y_0 = 2p/m$, a constant.

Hence the locus of the mid points of all parallel chords with slope m is the line parallel to the x axis and distant $2p/m$ from it, and the equation of this locus is

$$y = \frac{2p}{m}.$$

134. Diameter of a Conic. The locus of the mid points of any system of parallel chords of a conic is called a *diameter* of the conic.

The diameter of a circle is perpendicular to the chords bisected by it, but, in general, the perpendicularity is not true for other conics.

Exercise 39. Diameters

Find the equation of the diameter which bisects chords of the following parabolas under each of the specified conditions:

1. $y^2 = 12x$, chords parallel to $3x - y = 1$.

2. $y^2 = -x$, chords parallel to $x + y = 0$.

3. $3x = y^2$, chords perpendicular to $2x = 5y - 2$.

4. $y^2 = 6y + 4x - 1$, chords parallel to $2x = y$.

5. In the parabola $y^2 = 9x$ find the coordinates of the mid point of the chord $2x - 3y = 8$.

6. In the parabola $y^2 = 8x$ find the equation of the chord whose mid point is (6, 2).

7. In the parabola $y^2 = -10x$ what is the slope of the chords which are bisected by the line $y = -1$?

8. In the parabola $y^2 = 4px$, if the diameter which bisects all chords with slope m cuts the parabola in P, the tangent at P is parallel to these chords.

9. In the parabola $y^2 = 4px$ find the equation of the chord whose mid point is (a, b).

10. Having a parabola given, how do you find its axis? How do you find its focus?

11. Find the coordinates of two points A and B on the parabola $y^2 = 4x$, and prove that the tangents at A and B meet on the diameter which bisects AB.

12. A line l bisects all chords of slope m in the parabola $y^2 = 4px$. Find the slope of the chords bisected by l in the parabola $y^2 = 4qx$.

Exercise 40. Review

In the equation of the parabola $y^2 = 4px$ *find the value of* p, *given the following conditions:*

1. One end of the focal width is the point (6, 12).

2. The parabola passes through the point (−10, 6).

3. One end of the focal width is $5\sqrt{5}$ units from the vertex of the parabola.

4. For a point P on the parabola the subtangent is 15 and the focal radius is 10.

5. The line $3x - 4y = 12$ is tangent to the parabola.

6. The tangent to the parabola at a point whose abscissa is 6 has 2 for its y intercept.

7. Find the points on the parabola $y^2 = 12x$ which are 9 units from the focus.

8. Find the points on the parabola $y^2 = -24x$ which are 9 units from the vertex.

9. A parabolic trough is 4 ft. across the top and 2 ft. deep. Choose suitable axes and find the equation of the parabola.

10. If a parabolic reflector is 6 in. across the face and 5 in. deep, how far from the vertex of the parabola should the light be set that the rays may be reflected parallel to the axis?

11. This figure represents a parabolic arch, with $AB = 20$ ft., $CD = 6$ ft. Find the height of the arch at intervals of 2 ft. along AB.

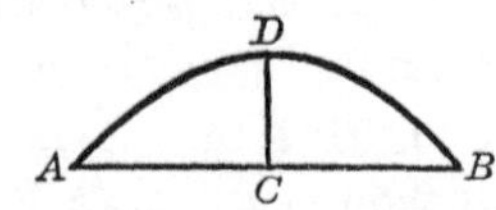

12. In the Brooklyn Bridge the distance AB between the towers is 1595 ft., the towers AP, BQ rise 158 ft. above the roadway AB, and PMQ is designed as a parabola. Find the length of an iron rod from PMQ perpendicular to AB, 100 ft. from A.

Assume AMB to be a straight line. This is not quite the case with this bridge. The cable PMQ would not form a parabola unless so weighted at regular intervals as to cause it to do so.

Find in each case the equation of the parabola subject to the following conditions:

13. The vertex is (4, 6) and the focus is (4, 0).

14. The focus is (2, – 3) and the directrix is $x = 8$.

15. The vertex is on the y axis, the axis is the line $y = 2$, and the focus is on the line $x + 2y = 7$.

16. The equation of a parabola with axis parallel to OX or parallel to OY involves three essential constants and may always be written in the form $y = ax^2 + bx + c$, when the axis is parallel to OY, or $x = ay^2 + by + c$, when the axis is parallel to OX.

17. What is the graph of the equation $Ay^2 + By + Cx + D = 0$ when $C = 0$?

18. Find the equation of the parabola through the points (2, 1), (5, – 2), and (10, 3), the axis being parallel to OX.

Represent the parabola by one of the forms given in §§ 121 and 123, or by one of those given in Ex. 16 above.

19. What parabola has the point (6, 5) for its vertex, has its axis parallel to OY, and passes through the point (10, 15)?

Find the vertex, focus, and focal width of each of the following parabolas, and draw the figure:

20. $x^2 + y = - 6$.

21. $3y^2 = 12y + 2x - 16$.

22. $x = y^2 + y + 1$.

23. $3x^2 + 8x - 7y = 16$.

Given the parabola $y^2 + Ay + Bx + C = 0$, find the following in terms of A, B, C:

24. The focus.

25. The vertex.

26. The focal width.

27. The directrix.

28. Show that the two parabolas $y^2 - 4y = 5x - 19$ and $5y = 1 + 6x - x^2$ are congruent. Draw the figures.

Prove that the distance from vertex to focus is the same for both.

29. What does the equation $y^2 = 4px$ become when the origin is moved to the focus?

30. Two congruent parabolas that have a common axis and a common focus but extend in opposite directions intersect at right angles.

31. Find the lowest point of the parabola $y = 3x^2 + 3x - 2$.

32. Find the highest point of the parabola $y = 2 - 3x - \frac{1}{2}x^2$.

33. What point of the parabola $x = 2y^2 - 4y + 5$ is nearest the y axis?

If a projectile starts with a velocity of 800 ft./sec. at an angle of 45° with the horizontal, find:

34. How high it goes, taking $g = 32$.

35. How far away, on a horizontal line, it falls.

This distance is called the *range* of the projectile. For present purposes ignore the resistance of the atmosphere.

If a projectile starts with a velocity of v ft./sec. at an angle of α degrees with the horizontal, prove that:

36. Its range is $v^2 \sin 2\alpha/g$, and is greatest when $\alpha = 45°$.

37. It reaches the maximum height $v^2 \sin^2\alpha/2g$.

38. The paths corresponding to various values of α all have the same directrix.

39. Find the area of the triangle formed with the axes by the tangent at the point $(-2, -4)$ to the parabola $y^2 = -8x$.

40. The slope of the tangent to the parabola $y^2 = 4px$ at the point whose ordinate is equal to the focal width is $\frac{1}{2}$.

41. Find the equations of the tangents to the parabola $y^2 = 4x$ at the points for which the focal radius is 10.

42. Find the sine and the cosine of the angle made with OX by the tangent at $(-9, 3)$ to the parabola $y^2 + x = 0$.

43. Find the equation of the normal at (x_1, y_1) to the parabola $y^2 = 4px$, and show that the x intercept is $x_1 + 2p$.

44. The angle between the tangents to the parabola $y^2 = 8x$ at the points $P_1(\frac{9}{2}, -6)$ and $P_2(8, 8)$ is equal to half the angle between the focal radii to P_1 and P_2.

45. The tangent to the parabola $6x = y^2$ at the point (x', y') passes through the point $(4, 7)$. Find x' and y'.

46. Find the tangents to the parabola $y^2 = 4x$ that pass through the point $(-12, 6)$.

47. A tangent to the parabola $y^2 = 20x$ is parallel to the line $x = y$. Find the point of contact.

48. Given that the line $y = 3x + k$ is tangent to the parabola $y^2 = 10x$, find the value of k.

49. No line $y = kx - k$, where k is real, can touch the parabola $x = y^2$.

50. The condition that the line $ax + by + c = 0$ shall be tangent to the parabola $y^2 = 4px$ is $ac = pb^2$.

51. Find the locus of the intersection of two perpendicular tangents to the parabola $y^2 = 4px$.

52. Given a parabola, find the locus of the intersection of two tangents whose inclinations are complementary.

53. A point P moves along the parabola $y^2 = 12x$ with a constant speed of 20 ft./sec. Find the horizontal and vertical components of the speed of P when P is at (3, ?); that is, find the speeds of P parallel to OX and OY.

Draw the rectangle with sides parallel to OX and OY, and with the diagonal PQ lying along the tangent, PQ representing 20, the speed.

54. A point starts at the vertex and moves on the parabola $y^2 = 8x$ with a constant horizontal speed of 16 ft./sec. Find its speed in its path, that is, in the direction of the tangent, at the end of 2 sec.; at the end of t seconds.

55. If A denotes the attraction between the sun and a comet C which moves on the parabola $y^2 = 4x$, the sun being at the focus, find the tangential and normal components of A when C is at the point (9, 6).

For the technical terms see Ex. 15, page 100.

56. The slope of the parabola $y = cx^2$ at (x_1, y_1) is $2\,cx_1$.

57. Using the result of Ex. 56, find the tangential and normal components of a weight of 1000 lb. attached to the parabola $x^2 = 16\,y$ at the point (16, 16).

58. The tangents to the parabola $y^2 = 4\,px$ at the points (h, k) and (h', k') intersect at the point $\left(\sqrt{hh'}, \frac{1}{2}[k + k']\right)$.

59. If the ordinates of P, Q, R on the parabola $y^2 = 4\,px$ are in geometric progression, the tangents at P and R meet on the ordinate of Q produced.

60. If a right triangle is inscribed in a parabola with the vertex of the right angle at the vertex of the parabola, then as the triangle turns about this vertex the hypotenuse turns about a fixed point on the axis of the parabola.

61. If a line l through the vertex cuts a parabola again in P, and if the perpendicular to l at P meets the axis in Q, the projection of PQ on the axis is constant.

62. Draw a parabola $y^2 = 4\,px$, and from any point Q on the focal width AB draw perpendiculars QR, QS to the tangents through A and B respectively. Find the coordinates of R and S, and prove that RS is tangent to the parabola.

63. Any circle with a focal radius of the parabola $y^2 = 4\,px$ as diameter is tangent to the y axis.

64. Any circle with a focal chord of the parabola $y^2 = 4\,px$ as diameter is tangent to the directrix.

65. Find the locus of the mid points of all ordinates of the parabola $y^2 = 4\,px$.

66. Find the locus of the mid points of all the focal radii of the parabola $y^2 = 4\,px$.

67. As a point P moves indefinitely far out on a parabola its distance from any line in the plane increases without limit.

68. Find the focus of the parabola which passes through the points (0, 6) and (3, 9) and has its axis along the y axis.

CHAPTER VIII

THE ELLIPSE

135. Ellipse. Given a fixed point F and a fixed line l, an *ellipse* is the locus of a point P which moves so that the ratio of its distances from F and l is a constant less than unity; that is, $FP/RP = e$ (§ 113).

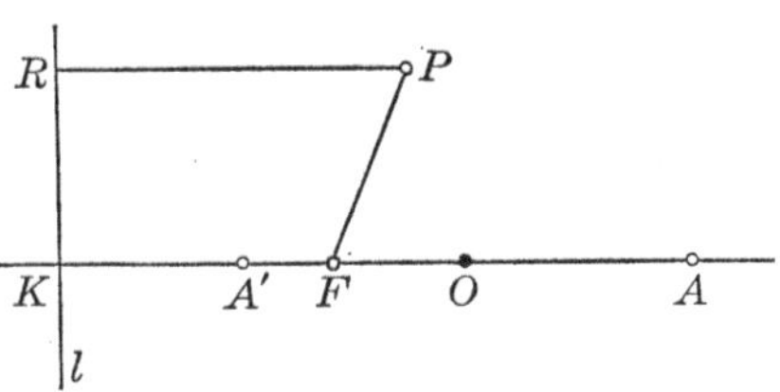

Drawing KF perpendicular to l, there is on KF a point A' so situated that $A'F/KA' = e$, and on KF produced a point A so situated that $FA/KA = e$. Therefore A' and A are on the ellipse. Now let $A'A = 2a$, and let O be the mid point of $A'A$, so that $A'O = OA = a$.

Let us now find KO and FO in terms of a and e.

Since $A'F = e \cdot KA'$, and $FA = e \cdot KA$, we have

$$A'F + FA = e(KA' + KA).$$

But $$A'F + FA = 2a,\ KA' = KO - a,$$

and $$KA = KO + a.$$

Then $$2a = e \cdot 2KO;$$

whence $$\mathbf{KO = \frac{a}{e}}.$$

Also, $$FA - A'F = e(KA - KA');$$

that is, $$(FO + a) - (a - FO) = e \cdot 2a;$$

whence $$\mathbf{FO = ae.}$$

Problem. Equation of the Ellipse

136. *To find the equation of the ellipse.*

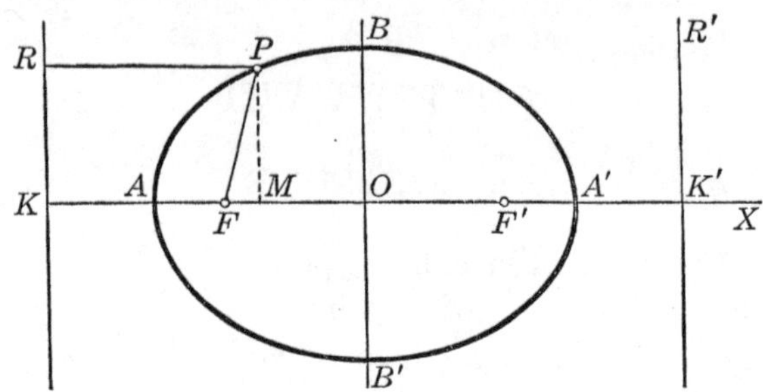

Solution. Taking the origin at O, the x axis perpendicular to the directrix, and the y axis parallel to the directrix, let $P(x, y)$ be any point on the ellipse.

Then the equation may be found from the condition

$$FP = e \cdot RP. \qquad \S\ 135$$

Since F is $(-ae, 0)$, then $FP = \sqrt{(x+ae)^2 + y^2}$. § 17

Since $$RP = KO - MO = \frac{a}{e} + x, \qquad \S\ 135$$

then $$e \cdot RP = e\left(\frac{a}{e} + x\right) = a + ex.$$

Therefore $$\sqrt{(x+ae)^2 + y^2} = a + ex,$$

or $$(1-e^2)x^2 + y^2 = a^2(1-e^2).$$

This equation of the ellipse, simple as it is, may be even more simply written by dividing both members by $a^2(1-e^2)$, and then, letting $a^2(1-e^2) = b^2$, we have

$$\frac{x^2}{a^2} + \frac{y^2}{b^2} = 1.$$

We shall often write this equation in the form $b^2x^2 + a^2y^2 = a^2b^2$. It may also be written $kx^2 + ly^2 = 1$, where $k = 1/a^2$ and $l = 1/b^2$.

137. Drawing the Ellipse. Solving the general equation, which was found in § 136, we have

$$y = \pm \frac{b}{a}\sqrt{a^2 - x^2}.$$

From this result we may easily prove that:

1. The x intercepts are a and $-a$.
2. The y intercepts are b and $-b$.

Since in the solution of the problem in § 136 we put b^2 for $a^2(1-e^2)$, and since $e<1$ by § 113, we see that $b<a$.

3. The value of y is real when $-a \leqq x \leqq a$, and imaginary in all other cases.
4. The curve is symmetric with respect to the coordinate axes OX and OY, and with respect to the origin O.

Let us now trace the ellipse in the first quadrant. As x increases from 0 to a, y decreases from b to 0, decreasing slowly when x is near 0, but decreasing rapidly when x is near a. These facts are represented by the curve BA', which we may now duplicate symmetrically in the four quadrants to make the complete ellipse.

For the plotting in a numerical case see § 42.

We therefore see that to draw an ellipse when the equation is in the form found in § 136 we may first find the intercepts $\pm a$ and $\pm b$, and then sketch the curve.

138. Second Focus and Directrix. If the figure shown in § 136 is revolved through 180° about the y axis, the ellipse revolves into itself, being symmetric with respect to the y axis. The focus then falls on OX at F', making $OF' = FO$, and the directrix KR falls in the position $K'R'$. Hence the ellipse may equally well be defined from the focus and directrix in the new positions, and we say that the ellipse has two foci, F and F', and two directrices.

139. Axes, Vertices, and Center. The segments $A'A$ and $B'B$ are called respectively the *major axis* and the *minor axis* of the ellipse. The ends A' and A of the major axis are called the *vertices* of the ellipse; the point O is called the *center* of the ellipse; and the segments OA and OB, or a and b, are called the *semiaxes* of the ellipse.

140. Eccentricity. The eccentricity e is related to the semiaxes a and b by the formula $b^2 = a^2(1 - e^2)$. § 136

Hence $$e = \frac{\sqrt{a^2 - b^2}}{a}.$$

It is, therefore, evident that the distance ae from focus to center (§ 135) is $FO = ae = \sqrt{a^2 - b^2}$.

141. Focal Width. The focal width or latus rectum (§ 120) of the ellipse $x^2/a^2 + y^2/b^2 = 1$ is found by doubling the positive ordinate at the focus; that is, the ordinate corresponding to $x = \sqrt{a^2 - b^2}$ (§ 140). Substituting this value of x, we have for the positive ordinate $y = b^2/a$.

Hence $$\textbf{focal width} = \frac{2b^2}{a}.$$

142. Major Axis as y Axis. If the y axis is taken along the major axis, that is, along the axis on which the foci lie, the equation of the ellipse is obviously, as before,

$$\frac{x^2}{a^2} + \frac{y^2}{b^2} = 1.$$

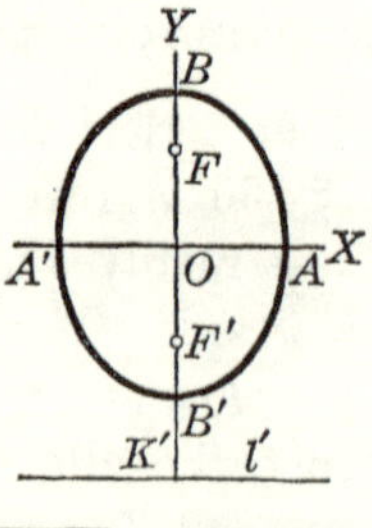

But in this case a is the minor semiaxis and b is the major semiaxis, so that a and b change places in the preceding formulas.

We therefore see that

$$a^2 = b^2(1 - e^2),\ e = \sqrt{b^2 - a^2}/b,\ F'O = be = \sqrt{b^2 - a^2},\ K'O = b/e.$$

THEOREM. SUM OF THE FOCAL RADII

143. *The sum of the distances from the foci to any point on an ellipse is constant and equal to the major axis.*

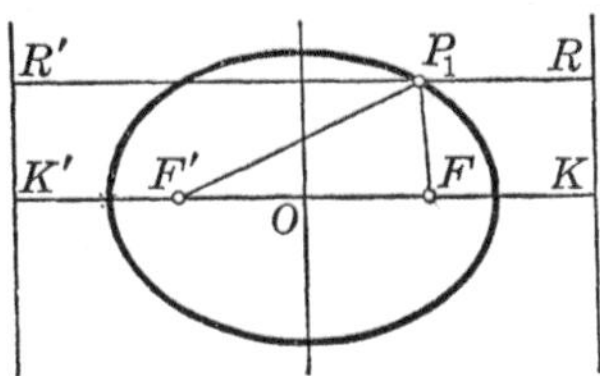

Proof. Let $P_1(x_1, y_1)$ be any point on the ellipse, F and F' the foci, and KR and $K'R'$ the directrices. By § 135,

$$FP_1 = e \cdot P_1R = e(OK - x_1) = e\left(\frac{a}{e} - x_1\right) = a - ex_1,$$

and $$F'P_1 = e \cdot R'P_1 = e(OK' + x_1) = e\left(\frac{a}{e} + x_1\right) = a + ex_1.$$

Therefore $$FP_1 + F'P_1 = 2a,$$

which, since a represents a constant, proves the theorem.

Hence an ellipse may be drawn by holding a moving pencil point P_1 against a string, the ends of which are fastened at F and F'.

144. Illustrative Examples. **1.** Show that the equation $4x^2 + 9y^2 = 36$ represents an ellipse, and find the foci, eccentricity, directrices, and focal width.

The equation, written in the form $x^2/9 + y^2/4 = 1$, represents an ellipse (§ 136) with $a^2 = 9$ and $b^2 = 4$. Using $a = 3$ and $b = 2$, the foci (§ 140) are $(\pm\sqrt{5}, 0)$; the eccentricity (§ 140) is $\frac{1}{3}\sqrt{5}$; the directrices (§ 135) are $x = \pm 9/\sqrt{5}$; and the focal width (§ 141) is $2\frac{2}{3}$.

2. Find the equation of the ellipse with the foci $(\pm 3, 0)$ and having $e = \frac{3}{5}$.

From the equations $\sqrt{a^2 - b^2} = 3$ and $\sqrt{a^2 - b^2}/a = \frac{3}{5}$ we have $a = 5$ and $b = 4$. Hence the required equation is $x^2/25 + y^2/16 = 1$.

Exercise 41. Equations of Ellipses

Draw the ellipse for each of the following equations and find the foci, eccentricity, directrices, and focal width:

1. $4x^2 + 25y^2 = 100$.

2. $9x^2 + 16y^2 = 144$.

3. $3x^2 + 4y^2 = 12$.

4. $3x^2 + 4y^2 = 24$.

5. $25x^2 + 9y^2 = 225$.

6. $9x^2 + 4y^2 = 36$.

7. $x^2 + 25y^2 = 25$.

8. $6x^2 + 9y^2 = 28$.

9. $2x^2 = 1 - y^2$.

10. $8x^2 + 3y^2 = 75$.

11. The equation $4x^2 + 9y^2 = -36$ can be written in the form of the equation of an ellipse, $x^2/a^2 + y^2/b^2 = 1$, but there are no real points on the locus.

12. The equation $Ax^2 + By^2 = C$, where A and B are positive, represents an ellipse that is real if C is positive, a single point if $C = 0$, and imaginary if C is negative.

13. Find in terms of A, B, and C the focus of the ellipse $Ax^2 + By^2 = C$ where A and B are positive and $A < B$.

Given the following conditions, find the equations of the ellipses having axes on XX' and YY', and draw the figures:

14. Foci, $(\pm 4, 0)$; vertices, $(\pm 6, 0)$.

15. Foci, $(\pm 3, 0)$; directrices, $x = \pm 12$.

16. Minor axis, 6; foci, $(\pm 4, 0)$.

17. Vertices, $(\pm 8, 0)$; eccentricity, $\frac{3}{4}$.

18. Vertices, $(0, \pm 8)$; eccentricity, $\frac{3}{4}$.

19. Eccentricity, $\frac{1}{2}$; major axis, 12; foci on OY.

20. Vertices, $(\pm 7, 0)$; the ellipse passes through $(1, 1)$.

21. Eccentricity, $\frac{3}{5}$; focal width, $\frac{32}{5}$.

22. The points $(2, 1)$ and $\left(1, \sqrt{\frac{19}{7}}\right)$ are on the ellipse.

The form $kx^2 + ly^2 = 1$, given in § 136, is the simplest for this case.

23. Find the equation of the ellipse if the distance between the foci equals the minor axis and the focal width is 4.

24. The line joining one end of the minor axis of an ellipse to one of the foci is equal to half the major axis.

25. An arch in the form of half an ellipse is 40 ft. wide and 15 ft. high at the center. Find the height of the arch at intervals of 10 ft. along its width.

26. An ellipse in which the major axis is equal to the minor axis is a circle.

27. In an ellipse with the fixed major axis $2a$ and the variable minor axis $2b$, as b approaches a the directrix moves indefinitely far away, the eccentricity approaches 0, and the foci approach the center.

28. Draw an ellipse having the major axis $2a$ and the minor axis $2b$. Describe a circle having the major axis of the ellipse as diameter. Taking any abscissa $OM = x$, draw the corresponding ordinates of the points P and P' on the ellipse and circle respectively. Show that the ordinates MP and MP' are in the constant ratio b/a.

29. Draw an ellipse, and then draw a circle on its major axis as diameter. Draw rectangles as in this figure. From Ex. 28 show that each rectangle R with a vertex on the ellipse and the corresponding rectangle R' with a vertex on the circle are related thus: $R = bR'/a$. Then, increasing indefinitely the number of rectangles, show that the area of the ellipse is πab.

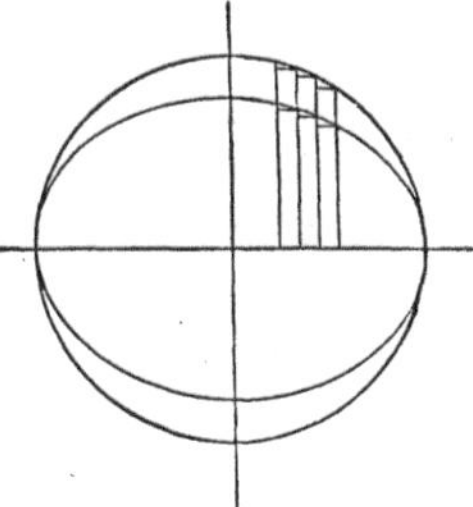

It should be observed that a circle is a special kind of ellipse, that in which $a = b = r$. In the circle $\pi ab = \pi aa = \pi r^2$.

30. Prove the converse of the theorem of § 143; that is, prove that the locus of a point which moves so that the sum of its distances from two fixed points F' and F is a constant $2a$ is an ellipse of which the foci are F' and F, and of which the major axis is $2a$.

PROBLEM. ELLIPSE WITH A GIVEN CENTER

145. *To find the equation of the ellipse having its center at (h, k) and its axes parallel to the coordinate axes.*

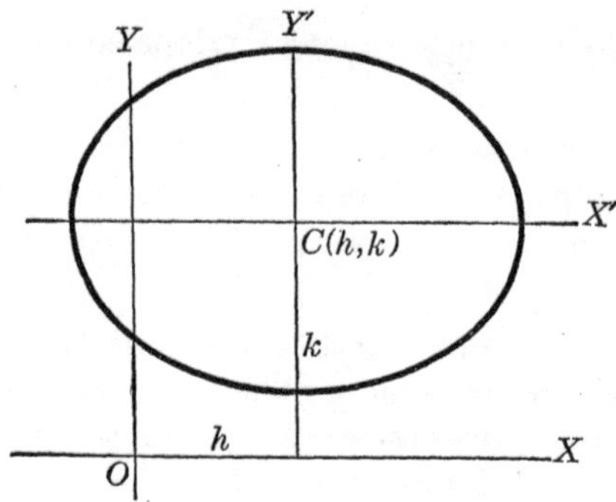

Solution. The equation of the ellipse referred to the axes CX' and CY', taken along the axes of the ellipse, is

$$\frac{x^2}{a^2}+\frac{y^2}{b^2}=1. \qquad \text{§ 136}$$

Now make a translation of the axes CX' and CY', moving the origin C to the point O, which with respect to C as origin is the point $(-h, -k)$. To effect this (§ 105) we write $x-h$ for x and $y-k$ for y, and the new equation of the ellipse, referred to OX and OY, is

$$\frac{(x-h)^2}{a^2}+\frac{(y-k)^2}{b^2}=1.$$

It should be observed that this equation, when cleared of fractions, is of the form $Ax^2+By^2+Cx+Dy+E=0$.

If a is greater than b, the major axis of the ellipse is parallel to OX; but if a is less than b, the minor axis is parallel to OX.

The form of the equation of an ellipse whose axes are not parallel to the coordinate axes will be considered in Chapter X.

THEOREM. THE EQUATION $Ax^2 + By^2 + Cx + Dy + E = 0$

146. *If A and B have the same sign and are not 0, every equation of the form $Ax^2 + By^2 + Cx + Dy + E = 0$ represents an ellipse having its axes parallel to the coordinate axes.*

Proof. By the process of completing squares such an equation may always be written in the form

$$A(x-h)^2 + B(y-k)^2 = G,$$

where h, k, and G are constants.

But this equation may be written

$$\frac{A(x-h)^2}{G} + \frac{B(y-k)^2}{G} = 1,$$

or

$$\frac{(x-h)^2}{G/A} + \frac{(y-k)^2}{G/B} = 1,$$

which, by § 145, represents an ellipse having its center at the point (h, k) and its semiaxes a and b such that $a^2 = G/A$ and $b^2 = G/B$.

For example, from the equation

$$4x^2 + 9y^2 - 16x + 18y = 11$$

we have

$$4(x^2 - 4x) + 9(y^2 + 2y) = 11,$$

or

$$4(x-2)^2 + 9(y+1)^2 = 36.$$

Then

$$\frac{(x-2)^2}{9} + \frac{(y+1)^2}{4} = 1.$$

When the equation $Ax^2 + By^2 + Cx + Dy + E = 0$ is reduced to the form $A(x-h)^2 + B(y-k)^2 = G$, if G has the same sign as A and B, the ellipse is obviously real; that is, real values of x and y satisfy the equation. If $G = 0$, the graph consists of one point only, the point (h, k), and is called a *point ellipse.* If G differs in sign from A and B, there are no real points on the graph, and the equation represents an *imaginary ellipse.* In this book we shall not consider imaginary conics.

Exercise 42. Axes Parallel to OX and OY

Draw the following ellipses, and find the center, eccentricity, and foci of each:

1. $x^2 + 4y^2 - 6x - 24y + 41 = 0.$

2. $4x^2 + 9y^2 + 16x - 18y - 11 = 0.$

3. $4x^2 + 25y^2 - 8x - 100y + 4 = 0.$

4. $2x^2 + 5y^2 - 16x + 20y + 42 = 0.$

5. $25x^2 + 4y^2 + 50x - 8y = 171.$

6. $2x^2 + 3y^2 + 12x - 12y = 6.$

7. $9x^2 + y^2 - 4y = 5.$

8. $9x^2 + 16y^2 - 12x + 16y = 64.$

9. $3x^2 + 7y^2 - 4x + y = 20.$

10. $x^2 + 2y^2 - 6y = 9.75.$

11. The equation $Ax^2 + By^2 + Cx + Dy + E = 0$, representing an ellipse, involves four essential constants and can always be written in the form $x^2 + cy^2 + dx + ey + f = 0.$

In finding the equation of an ellipse which satisfies the given conditions in a problem, it is well to make a choice among this form of the equation, the forms given in § 136, and the form given in § 145.

In each of the following cases find the equation of the ellipse having its axes parallel to OX and OY and fulfilling the given conditions, and draw the figure:

12. Center (4, 3), eccentricity $\frac{1}{2}$, and major axis parallel to OX, 12.

13. Foci (6, – 2) and (– 2, – 2), and major axis equal to twice the minor axis.

14. Center (1, 2), and passing through (1, 1) and (3, 2).

15. Vertices (0, – 2) and (0, 10), and one focus (0, 0).

16. Intercepts 2 and 8 on OX, and 2 and – 4 on OY.

PROBLEM. SLOPE OF THE ELLIPSE

147. *To find the slope of the ellipse $x^2/a^2 + y^2/b^2 = 1$ at any point $P_1(x_1, y_1)$ on the ellipse.*

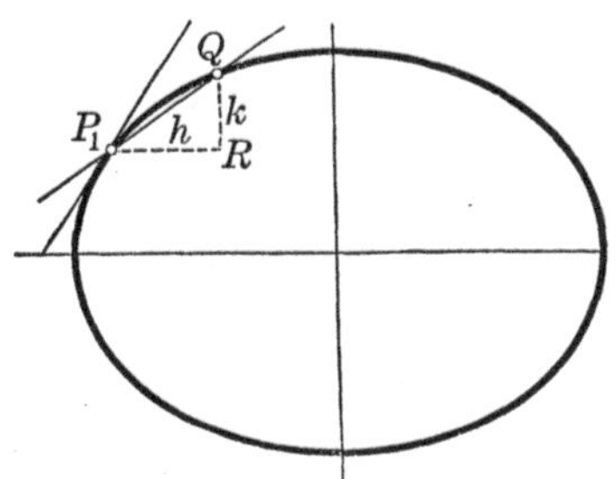

Solution. Let a second point on the given ellipse be $Q(x_1 + h, y_1 + k)$, where $PR = h$, $RQ = k$.

Then the slope of the secant P_1Q is k/h.

Since the points $P_1(x_1, y_1)$ and $Q(x_1 + h, y_1 + k)$ are on the ellipse $b^2x^2 + a^2y^2 = a^2b^2$, we have

$$b^2x_1^2 + a^2y_1^2 = a^2b^2, \tag{1}$$

and

$$b^2(x_1 + h)^2 + a^2(y_1 + k)^2 = a^2b^2. \tag{2}$$

Subtracting (1) from (2), we have

$$k(2a^2y_1 + a^2k) = -h(2b^2x_1 + b^2h);$$

whence

$$\frac{k}{h} = -\frac{2b^2x_1 + b^2h}{2a^2y_1 + a^2k}.$$

Now when $Q \rightarrow P_1$, $h \rightarrow 0$ and $k \rightarrow 0$; and hence

$$\lim k/h = -2b^2x_1/2a^2y_1.$$

But $\lim k/h = m$, the slope of the tangent at $P_1(x_1, y_1)$.

Hence

$$m = -\frac{b^2x_1}{a^2y_1},$$

which is also the slope of the ellipse at $P_1(x_1, y_1)$. § 125

Problem. Tangent to the Ellipse

148. *To find the equation of the tangent to the ellipse* $x^2/a^2 + y^2/b^2 = 1$ *at the point* $P_1(x_1, y_1)$.

Solution. The slope of the tangent at P_1 being $-\dfrac{b^2x_1}{a^2y_1}$ (§ 147), the equation of the tangent is

$$y - y_1 = -\frac{b^2x_1}{a^2y_1}(x - x_1); \qquad \text{§ 69}$$

that is,
$$b^2x_1x + a^2y_1y = b^2x_1^2 + a^2y_1^2.$$

Since (x_1, y_1) is on the ellipse, we have $b^2x_1^2 + a^2y_1^2 = a^2b^2$.

Hence
$$b^2x_1x + a^2y_1y = a^2b^2,$$

or
$$\frac{x_1x}{a^2} + \frac{y_1y}{b^2} = 1.$$

The equation of the tangent, therefore, may be obtained from the equation of the ellipse by writing x_1x for x^2, and y_1y for y^2.

149. Corollary 1. *The equation of the normal to the ellipse* $x^2/a^2 + y^2/b^2 = 1$ *at the point* $P_1(x_1, y_1)$ *is*

$$y - y_1 = \frac{a^2y_1}{b^2x_1}(x - x_1).$$

150. Corollary 2. *The intercepts of the tangent and normal at the point* $P_1(x_1, y_1)$ *on an ellipse are as follows:*

1. x intercept of tangent, $x = \dfrac{a^2}{x_1}$;
2. y intercept of tangent, $y = \dfrac{b^2}{y_1}$;
3. x intercept of normal, $x = \dfrac{a^2 - b^2}{a^2}x_1 = e^2x_1$; § 140
4. y intercept of normal, $y = \dfrac{b^2 - a^2}{b^2}y_1$.

Problem. Tangents having a Given Slope

151. *To find the equations of the lines which have the slope m and are tangent to the ellipse $x^2/a^2 + y^2/b^2 = 1$.*

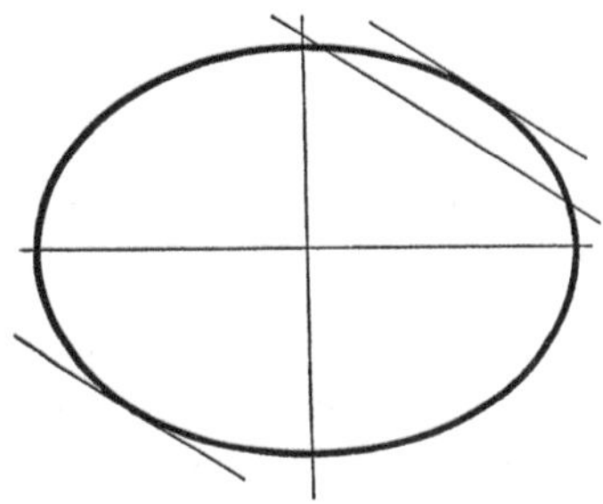

Solution. Let the equation $y = mx + k$ represent any line having the slope m. To find the common points of this line and the ellipse we regard their equations as simultaneous. Thus, the equation of the ellipse being

$$b^2x^2 + a^2y^2 = a^2b^2,$$

we have
$$b^2x^2 + a^2(mx + k)^2 = a^2b^2,$$

or
$$(b^2 + a^2m^2)x^2 + 2a^2mkx + a^2(k^2 - b^2) = 0.$$

The roots of this quadratic in x are the abscissas of the common points.

The condition under which the line $y = mx + k$ is tangent to the ellipse is that the common points coincide, and thus have the same abscissa; that is, that the roots of the above quadratic in x are equal, which requires that

$$(2a^2mk)^2 - 4(b^2 + a^2m^2)a^2(k^2 - b^2) = 0;$$

whence
$$k = \pm\sqrt{a^2m^2 + b^2}.$$

Hence there are two tangents having the slope m, namely,

$$\boldsymbol{y = mx \pm \sqrt{a^2m^2 + b^2}}.$$

Exercise 43. Tangents and Normals

In each of the following equations show that P is on the ellipse, find the tangent and normal at P, and draw the figure:

1. $3x^2+8y^2=35$; $P(1, 2)$.
4. $9y^2+x^2=25$; $P(4, -1)$.
2. $5x^2+2y^2=98$; $P(4, 3)$.
5. $9x^2+y^2=25$; $P(-1, -4)$.
3. $x^2+4y^2=25$; $P(3, -2)$.
6. $6x^2+11y^2=98$; $P(3, -2)$.

7. Find the equations of those tangents to the ellipse $7x^2+3y^2=28$ which have the slope $\frac{2}{3}$, and also find the point of contact and the intercepts of each tangent.

8. Draw the ellipse $x^2+25y^2=169$, the tangent at the point $P(12, 1)$, and two other tangents perpendicular to that tangent. Find the equations of the tangents.

9. Find the equation of that tangent to the ellipse $3x^2+4y^2=72$ which forms with the axes a triangle of area 21.

The student should find that there are eight such tangents.

10. If the normal at P to the ellipse $\frac{1}{2}x^2+2y^2=50$ passes through one end of the minor axis, find P.

It is obvious geometrically that the minor axis itself is such a normal, but there may be others. Let P be the point (h, k).

11. Through any point (h, k) two tangents to the ellipse may be drawn.

Consider the case of two coincident tangents and the case of imaginary tangents. The student may refer to property 5, § 130.

12. Find the equations of the tangents through the point $(8, -1)$ to the ellipse $2x^2+5y^2=70$.

13. Find the equations of those tangents to the ellipse $16x^2+9y^2=144$ which have equal intercepts.

14. An arch in the form of half an ellipse has a span of 40 ft. and a height of 15 ft. at the center. Draw the normal at the point of the ellipse which is 10 ft. above the major axis, and find where the normal cuts the major axis.

Find the distance to a focus of the ellipse $b^2x^2 + a^2y^2 = a^2b^2$ from a tangent drawn as follows:

15. At any point (x_1, y_1) on the ellipse.

16. At one end of a focal width.

17. Having the slope m.

Find the distance to the center of the ellipse $b^2x^2 + a^2y^2 = a^2b^2$ from a tangent drawn as follows:

18. At one end of a focal width.

19. Having the slope m.

20. Parallel to the line passing through one focus and one end of the focal width through the other focus.

21. Parallel to the line passing through one vertex and one end of the minor axis.

22. Using the proper formula in § 150, construct the tangent to the ellipse $x^2/a^2 + y^2/b^2 = 1$ at the point (x_1, y_1).

23. If we regard the major semiaxis a as constant, but regard b as having various values, then the equation $x^2/a^2 + y^2/b^2 = 1$ represents many ellipses, one for each value of b. Show that all the tangents to these ellipses at points which have the same abscissa meet on the x axis.

24. By Ex. 23 construct the tangents to a given ellipse from any point on the major axis produced.

25. No normal to an ellipse, excepting the major axis, passes between a focus and the vertex nearer that focus.

26. The product of the x intercepts of the tangent and normal at the point (x_1, y_1) is the constant $a^2 - b^2$.

27. The product of the y intercepts of the tangent and normal at the point (x_1, y_1) is the constant $b^2 - a^2$.

28. The tangent and normal at the point $P(x_1, y_1)$ bisect the angles between the focal radii of P.

Use the slopes of the lines concerned, and thus find the angles.

THEOREM. PROPERTY OF A NORMAL

152. *The normal at any point $P_1(x_1, y_1)$ on an ellipse bisects the angle between the focal radii of the point P_1.*

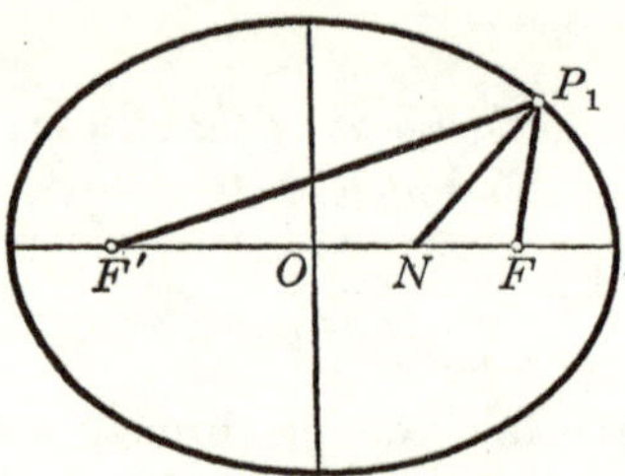

Proof. Since $$F'O = OF = ae, \qquad §§ 138, 135$$
and the x intercept of the normal is ON, where
$$ON = e^2x_1, \qquad § 150$$
we have
$$\frac{F'N}{NF} = \frac{F'O + ON}{OF - ON}$$
$$= \frac{ae + e^2x_1}{ae - e^2x_1}$$
$$= \frac{a + ex_1}{a - ex_1}.$$

Since in proving the theorem of § 143 we showed that $F'P_1 = a + ex_1$ and $FP_1 = a - ex_1$, we have
$$\frac{F'N}{NF} = \frac{F'P_1}{FP_1}.$$

Therefore, since P_1N divides $F'F$ into segments proportional to $F'P_1$ and FP_1, P_1N bisects the angle $F'P_1F$.

This proves an interesting property of the ellipse in regard to reflection. If a source of light, sound, or heat is at one focus, the waves are all reflected from the ellipse to the other focus.

Exercise 44. Properties of the Ellipse

1. The axes of an ellipse are normal to the ellipse, and no other normal passes through the center.

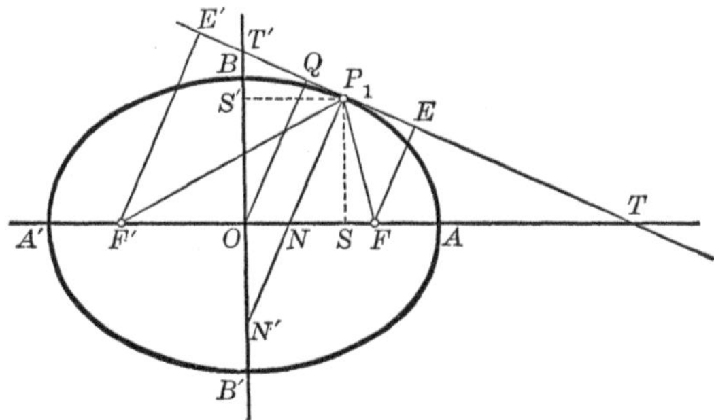

In the above figure, F and F' are the foci of the ellipse; P_1T is the tangent at $P_1(x_1, y_1)$, the slope of this tangent being m; P_1N is the normal at P_1; F'E', OQ, and FE are perpendicular to P_1T; and P_1S and P_1S' are perpendicular to the axes. Prove the following properties of the ellipse:

2. $ST = (a^2 - x_1^2)/x_1$.

3. $S'T' = (b^2 - y_1^2)/y_1$.

4. $NS = b^2x_1/a^2$.

5. $N'S' = a^2y_1/b^2$.

6. $ON \cdot OT = \overline{OF}^2$.

7. $OS \cdot OT = \overline{OA}^2 = a^2$.

8. $OQ \cdot NP_1 = \overline{OB}^2 = b^2$.

9. $OQ \cdot N'P_1 = a^2$.

10. $N'P_1 \cdot NP_1 = F'P_1 \cdot FP_1$.

11. $N'P_1 \cdot NP_1 = T'P_1 \cdot P_1T$.

12. $FE \cdot F'E' = b^2$.

13. $a^2 \cdot \overline{SP_1}^2 = b^2 \cdot A'S \cdot SA$.

14. As the tangent turns about the ellipse the locus of E, the foot of the perpendicular from F to the tangent, is the circle having the center O and the radius a.

A simple method is to represent the tangent P_1T by the equation $y = mx + \sqrt{a^2m^2 + b^2}$ and the line through $F(ae, 0)$ perpendicular to P_1T by the equation $y = -(x - ae)/m$, or $x + my = ae$; then, regarding these equations as simultaneous, eliminate m and obtain the equation $x^2 + y^2 = a^2$, which is the circle with center O and radius a.

AG

Problem. Locus of Mid Points of Parallel Chords

153. *To find the locus of the mid points of a system of parallel chords of an ellipse.*

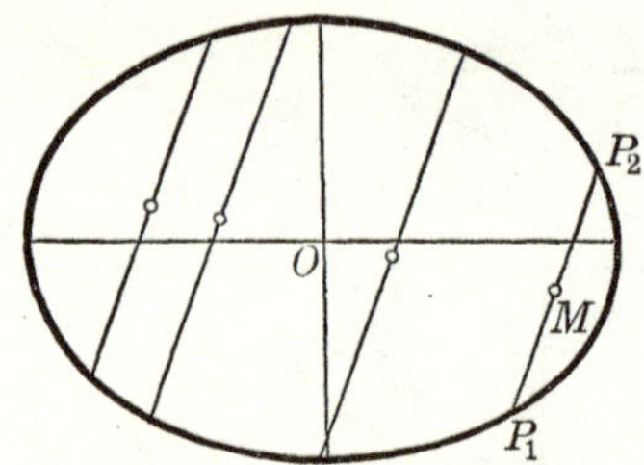

Solution. Let the ellipse be represented by the equation $b^2x^2 + a^2y^2 = a^2b^2$; the slope of the parallel chords by m; any one of the chords, say P_1P_2, by $y = mx + k$; and the mid point of this chord by $M(x', y')$.

For the coordinates of P_1 and P_2 we regard the equations $b^2x^2 + a^2y^2 = a^2b^2$ and $y = mx + k$ as simultaneous.

Eliminating y, we have

$$b^2x^2 + a^2(mx + k)^2 = a^2b^2;$$

that is, $$(a^2m^2 + b^2)x^2 + 2\,a^2mkx + a^2k^2 - a^2b^2 = 0.$$

The roots of this quadratic in x are the abscissas x_1 and x_2 of the common points P_1 and P_2, and the abscissa of the mid point M is half their sum. That is, since the sum of the roots x_1 and x_2 is $-\dfrac{2\,a^2mk}{a^2m^2 + b^2}$, we have

$$x' = \tfrac{1}{2}(x_1 + x_2) = -\frac{a^2mk}{a^2m^2 + b^2}. \qquad (1)$$

If the student does not recall the fact that the sum of the roots of the general quadratic $Ax^2 + Bx^2 + C = 0$ is $-B/A$, he may consult § 2 on page 283 of the Supplement.

Since $M(x', y')$ is on the line $y = mx + k$, we have

$$y' = mx' + k. \tag{2}$$

We now have relations (1) and (2) involving x' and y', the constant slope m, and the y intercept k, which is different for different chords. If we combine these relations in such a way as to eliminate k, we obtain the desired relation between the coordinates x' and y' of the mid point of any chord. Substituting in (1), we have

$$x' = -\frac{a^2m(y' - mx')}{b^2 + a^2m^2};$$

that is,

$$y' = -\frac{b^2}{a^2m}x';$$

or, using x and y for the variable coordinates of M,

$$y = -\frac{b^2}{a^2m}x. \tag{3}$$

This is the equation of the diameter which bisects the chords having the slope m. It is obviously a straight line through the center of the ellipse, and it is evident that every straight line through the center is a diameter.

The above method fails if the chords are parallel to OY, for then the equation of any one of the chords is $x = k'$, and not $y = mx + k$. But chords parallel to OY are bisected by the major axis, since the ellipse is symmetric with respect to the x axis.

154. Corollary. *If m' is the slope of the diameter which bisects the chords of slope m, then*

$$m'm = -\frac{b^2}{a^2}.$$

Since the equation of the diameter is (3) above, its slope is $-b^2/a^2m$. That is, $m' = -b^2/a^2m$, and $m'm = -b^2/a^2$.

Theorem. Conjugate Diameters

155. *If one diameter of an ellipse bisects the chords parallel to another diameter, the second diameter bisects the chords parallel to the first.*

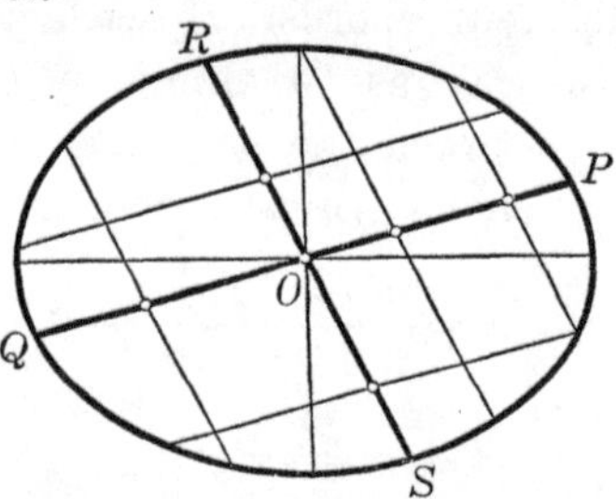

Proof. Let m' be the slope of the first diameter PQ, and m the slope of the second diameter RS.

Since by hypothesis PQ bisects the chords parallel to RS,

$$m'm = -b^2/a^2. \qquad \text{§ 154}$$

But this is also the condition under which RS bisects the chords parallel to PQ, and hence the theorem is proved.

156. Conjugate Diameters. If each of two diameters of an ellipse bisects the chords parallel to the other, the diameters are called *conjugate diameters*.

Exercise 45. Diameters

Given the ellipse $9x^2 + 16y^2 = 144$, find the equations of:

1. The diameter which bisects the chords having the slope $\frac{1}{2}$.

2. The chord of which the mid point is $(2, -1)$.

3. Two conjugate diameters, one of which passes through the point $(1, 2)$.

4. The chord which passes through the point $(6, 10)$ and is bisected by the diameter $x + 2y = 0$.

5. In Ex. 4 find the mid point of the chord $3x - 2y = 7$.

THEOREM. TANGENTS AT ENDS OF A DIAMETER

157. *The tangents at the ends of a diameter of an ellipse are parallel to each other and to the conjugate diameter.*

Proof. Let $P(x_1, y_1)$ and $Q(-x_1, -y_1)$ be the ends of any diameter PQ. Then the slope m of PQ is y_1/x_1. Hence m', the slope of the conjugate diameter RS, found from the relation $mm' = -b^2/a^2$ (§ 154) is $-b^2x_1/a^2y_1$. But the slopes of the tangents at $P(x_1, y_1)$ and $Q(-x_1, -y_1)$ are both $-b^2x_1/a^2y_1$ (§ 147), and hence these tangents are parallel to each other and to the conjugate diameter.

PROBLEM. ENDS OF A CONJUGATE DIAMETER

158. *Given one end $P(x_1, y_1)$ of a diameter of an ellipse, to find the ends of the conjugate diameter.*

Solution. Let PQ be the diameter through P and let RS be the conjugate diameter.

We may show, as in the proof of § 157, that the slope of RS is $-\dfrac{b^2x_1}{a^2y_1}$; then the equation of RS is $y = -\dfrac{b^2x_1}{a^2y_1}x$.

To find the coordinates of R and S, we regard this equation and the equation of the ellipse $b^2x^2 + a^2y^2 = a^2b^2$ as simultaneous. Eliminating y and simplifying, we have

$$x^2\frac{b^2x_1^2 + a^2y_1^2}{a^2y_1^2} = a^2.$$

But since $P(x_1, y_1)$ is on the ellipse, $b^2x_1^2 + a^2y_1^2 = a^2b^2$.

Hence $x^2 = \dfrac{a^2}{b^2}y_1^2$, and $x = \pm\dfrac{a}{b}y_1$, $y = -\dfrac{b^2x_1}{a^2y_1}x = \mp\dfrac{b}{a}x_1$,

and the ends of the conjugate diameter RS are

$$R\left(-\frac{a}{b}y_1, \frac{b}{a}x_1\right) \text{ and } S\left(\frac{a}{b}y_1, -\frac{b}{a}x_1\right).$$

THEOREM. ANGLE BETWEEN CONJUGATE DIAMETERS

159. *If θ is the angle from one semidiameter a' of an ellipse to its conjugate b', then $\sin\theta = ab/a'b'$.*

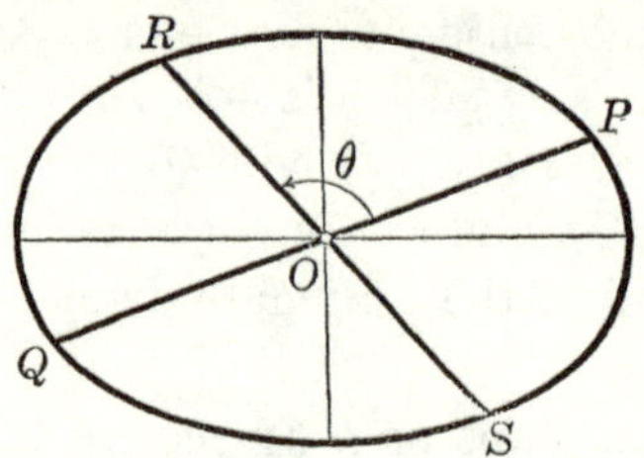

Proof. Let OP and OR be the semidiameters a' and b' with inclinations α and α', and let P be the point (x_1, y_1).

Then R is the point $(-ay_1/b, bx_1/a)$. § 158

Then $\sin\alpha = y_1/a'$, $\cos\alpha = x_1/a'$, $\sin\alpha' = \dfrac{bx_1/a}{b'} = \dfrac{b}{ab'}x_1$,

and $\cos\alpha' = -\dfrac{ay_1/b}{b'} = -\dfrac{a}{bb'}y_1$.

Since $\theta = \alpha' - \alpha$, $\sin\theta = \sin\alpha'\cos\alpha - \cos\alpha'\sin\alpha$, or

$$\sin\theta = \frac{b}{ab'}x_1 \cdot \frac{x_1}{a'} + \frac{a}{bb'}y_1 \cdot \frac{y_1}{a'} = \frac{b^2x_1^2 + a^2y_1^2}{aba'b'} = \frac{a^2b^2}{aba'b'} = \frac{ab}{a'b'}.$$

Exercise 46. Conjugate Diameters

1. In general, two conjugate diameters of an ellipse are not perpendicular to each other. State the exceptional case.

2. Two conjugate diameters of an ellipse cannot both lie within the same quadrant.

3. If a' and b' are any two conjugate semidiameters of an ellipse, then $a'^2 + b'^2 = a^2 + b^2$.

4. Find the equations of the four tangents at the extremities of two conjugate diameters of the ellipse $5x^2 + 2y^2 = 38$, one of the extremities being the point $(2, -3)$.

160. Auxiliary Circle. The circle having for diameter the major axis of the ellipse is called the *major auxiliary circle* of the ellipse; obviously, its equation is $x^2 + y^2 = a^2$.

If P is a point on the ellipse ABA', and the ordinate MP produced meets the major auxiliary circle in Q, the points P and Q are said to be *corresponding points* of the ellipse and the circle.

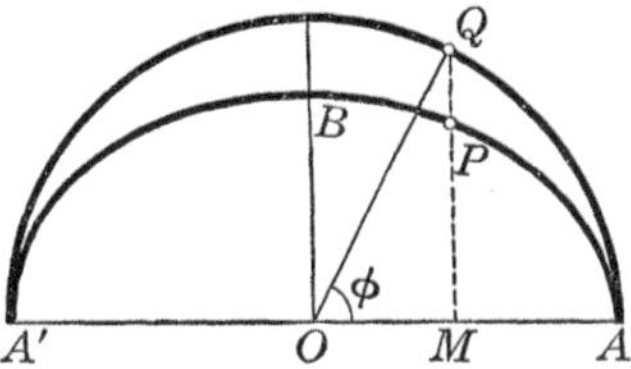

The circle $x^2 + y^2 = b^2$, which has for diameter the minor axis, is called the *minor auxiliary circle* of the ellipse.

161. Eccentric Angle. The angle MOQ is called the *eccentric angle* of the ellipse for the point P, and is denoted by the letter ϕ.

THEOREM. ELLIPSE AND MAJOR AUXILIARY CIRCLE

162. *The ordinates of corresponding points on the ellipse* $b^2x^2 + a^2y^2 = a^2b^2$ *and the major auxiliary circle are in the constant ratio* $b : a$.

Proof. Denote by x the common abscissa OM of P and Q. Then since $P(x, MP)$ is on the ellipse and $Q(x, MQ)$ is on the circle, we have

$$b^2x^2 + a^2\overline{MP}^2 = a^2b^2,$$

and
$$x^2 + \overline{MQ}^2 = a^2;$$

whence
$$MP = \pm \frac{b}{a}\sqrt{a^2 - x^2},$$

and
$$MQ = \pm \sqrt{a^2 - x^2}.$$

Obviously, then, we have the proportion

$$MP : MQ = b : a.$$

PROBLEM. ECCENTRIC ANGLE

163. *To express the coordinates of any point $P(x, y)$ on an ellipse in terms of the eccentric angle for the point P.*

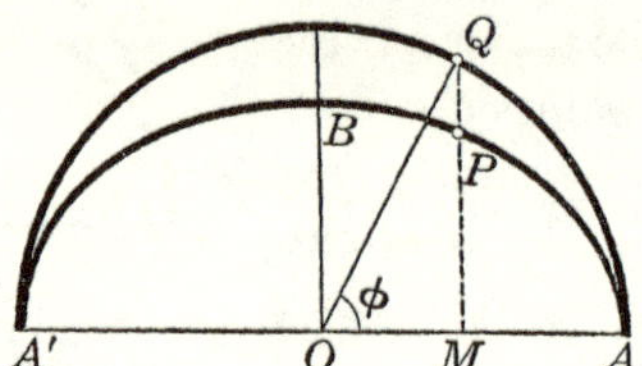

Proof. Since $OQ = a$, we have $x = a\cos\phi$.

And since $$MP = \frac{b}{a}\cdot MQ, \qquad \S\ 162$$

we have $$y = \frac{b}{a}\cdot MQ = \frac{b}{a}\cdot a\sin\phi.$$

That is, $$\boldsymbol{x = a\cos\phi,}$$

and $$\boldsymbol{y = b\sin\phi.}$$

These equations possess the advantage of expressing the variable coordinates x and y in terms of the single variable ϕ.

Exercise 47. Eccentric Angles

Find the eccentric angle for each point in Exs. 1 and 2:

1. The point $(2, \sqrt{3})$ on the ellipse $x^2 + 4y^2 = 16$.

2. The point (x_1, y_1) on the ellipse $b^2x^2 + a^2y^2 = a^2b^2$.

3. The tangent to the ellipse $b^2x^2 + a^2y^2 = a^2b^2$ at the point for which the eccentric angle is ϕ is $bx\cos\phi + ay\sin\phi = ab$.

4. If ϕ is the eccentric angle for an end of a diameter of the ellipse $b^2x^2 + a^2y^2 = a^2b^2$, the ends of the conjugate diameter are $(-a\sin\phi, b\cos\phi)$ and $(a\sin\phi, -b\cos\phi)$.

5. The eccentric angles for the ends of two conjugate semi-diameters differ by 90°.

Exercise 48. Review

1. The ellipses $4x^2+9y^2=36$ and $4x^2+9y^2=72$ have the same eccentricity.

2. The ellipses $16x^2+9y^2=144$ and $17x^2+10y^2=170$ have the same foci.

3. If the ellipse $4x^2+7y^2=36$ is rotated 90° about its center, it coincides with the ellipse $7x^2+4y^2=36$.

4. All the ellipses which are represented by the equation

$$\frac{x^2}{a^2}+\frac{y^2}{b^2}=k,$$

for various positive values of k, have the same center, have their axes in the same ratio, and have the same eccentricity.

5. Draw any two of the ellipses of Ex. 4 and a number of parallel lines cutting chords from both ellipses. Then the line which bisects the chords of one ellipse also bisects the chords of the other ellipse.

6. The two segments of any line intercepted between the two ellipses of Ex. 5 are equal, and any chord of the larger ellipse which is tangent to the smaller ellipse is bisected at the point of tangency.

7. All the ellipses which are represented by the equation

$$\frac{x^2}{a^2+k}+\frac{y^2}{b^2+k}=1,$$

for various values of k, have the same foci.

If k is negative and its absolute value lies between a^2 and b^2, either the x^2 term or the y^2 term is negative. In neither case does the equation represent an ellipse.

8. Draw the axes of an ellipse, having given the foci and one point on the curve.

9. Having given one point of an ellipse and the length and position of the major axis, draw the minor axis and the foci.

10. This figure represents an arch formed by less than half an ellipse. Chord AB is 20 ft. long and is 4 ft. from the highest point E. Chord CD is 16 ft. long and is 2 ft. from E. Find the height of the arch at intervals of 5 ft. along AB.

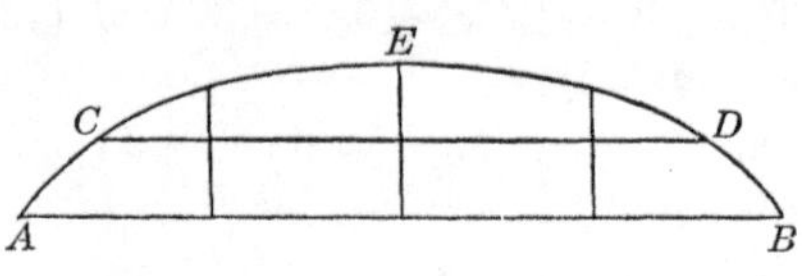

11. Find the ratio of the two axes of an ellipse if the center and foci divide the major axis into four equal parts.

Find the equations of the ellipses having axes on the coordinate axes and satisfying the following conditions:

12. Sum of axes, 54; distance between the foci, 36.

13. Major axis, 20; minor axis equal to the distance between the foci.

14. Sum of the focal radii of a point on the ellipse, three times the distance from focus to vertex; minor axis, 8.

15. Draw a circle and an ellipse having the same center, the diameter of the circle being less than the major axis and greater than the minor axis of the ellipse. Prove that the quadrilateral having for vertices the four common points of the circle and ellipse is a rectangle having its sides parallel to the axes. If this rectangle is a square and the axes of the ellipse are given, find the radius of the circle.

16. The minor semiaxis of an ellipse is the mean proportional between the segments of the major axis made by a focus.

17. Prove the theorem of Ex. 3, page 160, by using the result of Ex. 4, page 162, and that of § 163.

18. If an end P of a diameter of an ellipse (§ 163) is the point $(a \cos \phi, b \sin \phi)$, then, from Ex. 4, page 162, an end of the conjugate diameter is the point $R(-a \sin \phi, b \cos \phi)$. Write the equations of the tangents at P and R, and prove that these tangents intersect at the point whose coordinates are $a(\cos \phi - \sin \phi)$ and $b(\cos \phi + \sin \phi)$.

Find the equations of the ellipses having axes parallel to the coordinate axes, and satisfying the following conditions:

19. Origin at the left end of the major axis.

20. Origin at the right end of the major axis.

21. Origin at the upper end of the minor axis.

22. Origin at one focus.

23. Axes 10 and 16; center $(-3, 2)$.

24. Center $(-1, 1)$; one focus $(-1, 5)$; one end of major axis $(-1, -5)$.

25. Given an ellipse, find by construction its center, foci, and axes.

26. Show that the two ellipses $2x^2 + 3y^2 + 4x - 6y = 0$ and $2x^2 + 5y^2 + 4x - 10y = 0$ have the same center, and find the coordinates of their common points.

If the origin is first moved to the common center, the solution of the simultaneous equations is much simpler.

27. Find the ratio in which the abscissa of any point P on the ellipse $b^2x^2 + a^2y^2 = a^2b^2$ is divided by the normal at P.

28. Given an ellipse, find the locus of the intersection of tangents which are perpendicular to each other.

If $y = mx + \sqrt{a^2m^2 + b^2}$ represents one tangent, then the other is represented by $y = -\frac{1}{m}\cdot x + \sqrt{a^2\left(-\frac{1}{m}\right)^2 + b^2}$. See note to Ex. 14, page 155.

29. Find the locus of the intersection of a tangent to the ellipse and the perpendicular from the origin to the tangent.

This locus, obviously passing through the ends of the axes of the ellipse but elsewhere a little broader than the ellipse, is called an *oval*.

30. The tangents at the ends of any chord of an ellipse intersect on the diameter which bisects the chord.

31. Find the points at which tangents that are equally inclined to the axes touch the ellipse $b^2x^2 + a^2y^2 = a^2b^2$.

32. Find the condition that the line $x/m + y/n = 1$ is tangent to the ellipse $x^2/a^2 + y^2/b^2 = 1$.

33. The parallelogram which is formed by the four tangents at the ends of two conjugate diameters of an ellipse has a constant area.

Recall that the area of a parallelogram is $a'b' \sin \theta$, where a' and b' are adjacent sides and θ is the included angle, and then use § 159.

34. If the ends P and P' of any diameter are joined to any point Q on the ellipse $b^2x^2 + a^2y^2 = a^2b^2$, the diameters parallel to PQ and $P'Q$ are conjugate.

Prove that the product of the slopes of PQ and $P'Q$ is $-b^2/a^2$.

35. Draw the rectangle formed by the tangents at the ends of the axes of an ellipse. Prove that the diameters along the diagonals of the rectangle are conjugate and equal.

36. Find the eccentric angles for the ends of the equal conjugate diameters of Ex. 35.

37. The path of the earth is an ellipse, the sun being at one focus. Find the equation and eccentricity of the ellipse if the distances from the sun to the ends of the major axis are respectively 90 and 93 millions of miles.

38. Find the locus of the mid points of the ordinates of a circle that has its center at the origin.

39. Find the locus of the mid points of the chords drawn through one end of the minor axis of an ellipse.

40. The sum of the squares of the reciprocals of two perpendicular diameters of an ellipse is constant.

41. If the tangents at the vertices A' and A of an ellipse meet any other tangent in the points C' and C respectively, then $A'C' \cdot AC = b^2$.

42. Determine the number of normals from a given point to a given ellipse.

43. The circle having as diameter a focal radius of an ellipse is tangent to the major auxiliary circle.

CHAPTER IX

THE HYPERBOLA

164. Hyperbola. Given a fixed point F and a fixed line l, a *hyperbola* is the locus of a point P which moves so that the ratio of its distances from F and l is a constant greater than unity; that is, $FP/PR = e$ (§ 113).

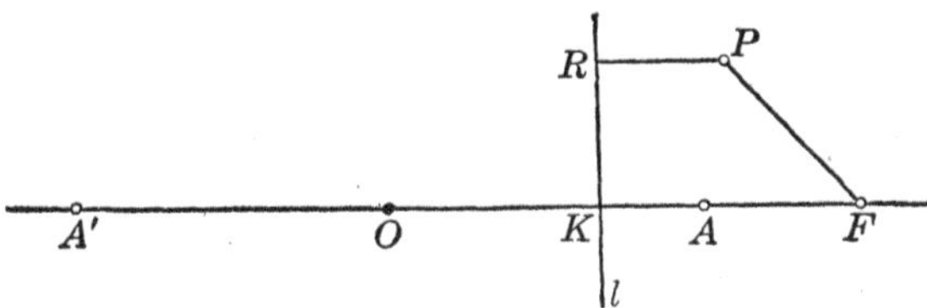

Draw KF perpendicular to l. Then on KF there is a point A such that $AF/KA = e$, and on FK produced there is a point A' such that $A'F/A'K = e$. That is, $AF = e \cdot KA$, and $A'F = e \cdot A'K$.

Then, by definition, A and A' are on the hyperbola. Now let $A'A = 2a$, let O be the mid point of $A'A$, so that $A'O = OA = a$, and find OK and OF in terms of a and e.

Since $\quad A'F - AF = e(A'K - KA)$,

that is, $\quad 2a = e(\overline{a + OK} - \overline{a - OK})$,

we have $\quad \boldsymbol{OK = \dfrac{a}{e}}.$

Also, $\quad A'F + AF = 2\,OF = e(A'K + KA) = e \cdot 2a$;

whence $\quad \boldsymbol{OF = ae}.$

Problem. Equation of the Hyperbola

165. *To find the equation of the hyperbola.*

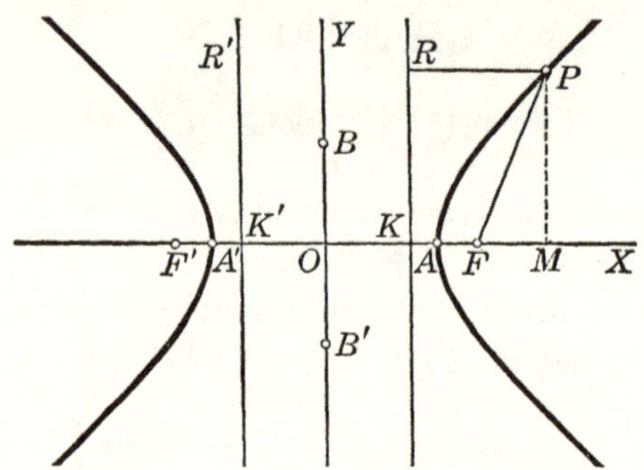

Solution. Taking the origin at O (§ 164), the x axis perpendicular to the directrix, and the y axis parallel to the directrix, let $P(x, y)$ be any point on the hyperbola.

Then the equation may be found from the condition

$$FP = e \cdot RP. \qquad \S\,164$$

Since F is $(ae, 0)$, then $FP = \sqrt{(x - ae)^2 + y^2}$. § 17

Since $RP = OM - OK = x - a/e,$ § 164

then $e \cdot RP = e(x - a/e) = ex - a.$

Therefore $ex - a = \sqrt{(x - ae)^2 + y^2};$

whence $(e^2 - 1)x^2 - y^2 = a^2(e^2 - 1),$

or $$\frac{x^2}{a^2} - \frac{y^2}{a^2(e^2 - 1)} = 1.$$

Letting the positive quantity $a^2(e^2 - 1) = b^2$, we have

$$\frac{x^2}{a^2} - \frac{y^2}{b^2} = 1.$$

This equation is often written $b^2x^2 - a^2y^2 = a^2b^2$. It may also be written $kx^2 - ly^2 = 1$, where $k = 1/a^2$ and $l = 1/b^2$. Compare § 136.

166. Shape of the Hyperbola. The shape of the hyperbola is easily inferred from the equation

$$\frac{x^2}{a^2}-\frac{y^2}{b^2}=1,$$

or from the derived equation

$$y=\pm\frac{b}{a}\sqrt{x^2-a^2}.$$

From this equation we may easily prove that:

1. The x intercepts are a and $-a$.
2. The y intercepts are imaginary.
3. The value of y is real when x is numerically equal to or greater than a, and is either positive or negative; but y is imaginary when $-a<x<a$.
4. The curve is symmetric with respect to the axes OX and OY, and also with respect to the origin O.

Let us trace the hyperbola in the first quadrant. When $x=a$, $y=0$; and as x increases without limit, y increases without limit. Duplicating the curve symmetrically in each of the other quadrants, we have the complete hyperbola.

167. Second Focus and Directrix. As in the case of the ellipse (§ 138), the hyperbola may equally well be defined from a second focus F' and a second directrix $K'R'$. It is evident that F' and F, and $K'R'$ and KR, are symmetrically located with respect to OY.

168. Axes, Vertices, and Center. The segment $A'A$ is called the *real axis* and the point O the *center* of the hyperbola. The points A' and A are called the *vertices*. Though the curve does not cut the y axis (§ 166, 2), we lay off $OB=b$, $OB'=-b$, and call $B'B$ the *conjugate axis*.

The real axis is also commonly called the *transverse axis*; but since the term is often confusing we shall use the simpler one given above, leaving it to the instructor to change to the older usage if desired.

169. Eccentricity. The eccentricity e is related to the semiaxes a and b by the formula $a^2(e^2-1)=b^2$ (§ 165).

Hence
$$e=\frac{\sqrt{a^2+b^2}}{a}.$$

It is, therefore, evident that the distance ae from center to focus (§ 164) is $OF=ae=\sqrt{a^2+b^2}$.

170. Focal Width. The focal width (§ 120) of the hyperbola $x^2/a^2-y^2/b^2=1$ is found by doubling the positive ordinate at the focus; that is, by doubling the ordinate corresponding to $x=\sqrt{a^2+b^2}$ (§ 169). This gives, as in § 141,

$$\text{focal width}=\frac{2b^2}{a}.$$

171. Real Axis as y Axis. If the y axis is taken along the real axis, that is, along the axis on which the foci lie, the x and y terms of § 165 exchange places in the equation, and we have

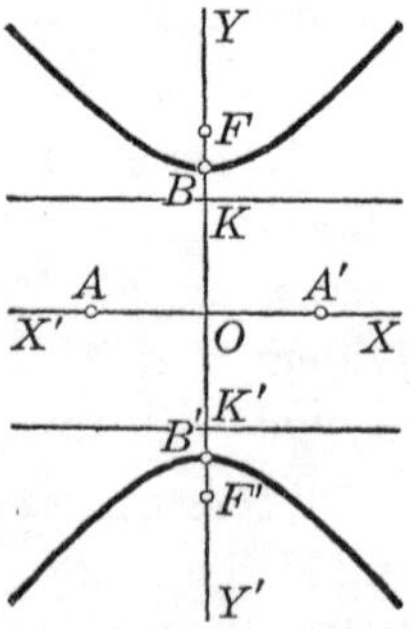

$$\frac{y^2}{b^2}-\frac{x^2}{a^2}=1,$$

where $2b$ denotes the real axis $B'B$, and $2a$ the conjugate axis $A'A$. Also we have the changed formulas

$$a^2=b^2(e^2-1),\quad e=\sqrt{a^2+b^2}/b,$$
$$OF=be=\sqrt{a^2+b^2},\quad OK=b/e.$$

The student should compare this work with that given in § 142.

172. Asymptote. It follows from the definition of an asymptote in § 46, that if any segment between a curve and a straight line approaches 0 when the segment moves indefinitely far away, the line is an asymptote to the curve.

THEOREM. ASYMPTOTES TO THE HYPERBOLA

173. *The lines* $y = \frac{b}{a}x$ *and* $y = -\frac{b}{a}x$ *are asymptotes to the hyperbola* $b^2x^2 - a^2y^2 = a^2b^2$.

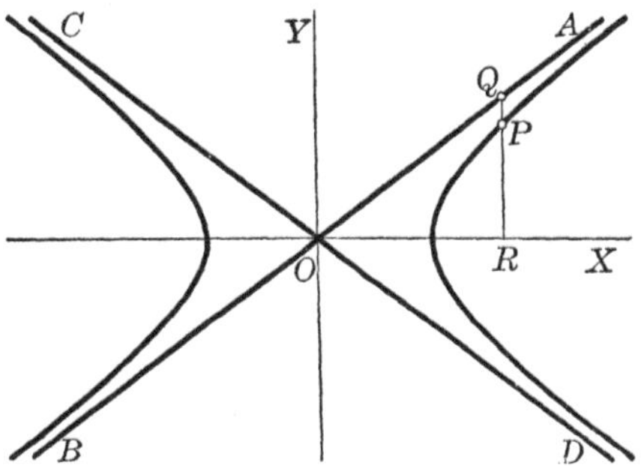

Proof. Let AB and CD represent the lines $y = \pm bx/a$ through O. Let P and Q be points on the hyperbola and AB respectively having the same abscissa h. We are to prove that $PQ \to 0$ when h increases without limit.

Since $Q(h, RQ)$ is on the line $y = bx/a$, then $RQ = bh/a$.

Since $P(h, RP)$ is on the hyperbola $b^2x^2 - a^2y^2 = a^2b^2$, then $b^2h^2 - a^2\overline{RP}^2 = a^2b^2$; whence $RP = b\sqrt{h^2 - a^2}/a$.

Hence $$PQ = RQ - RP = \frac{b}{a}(h - \sqrt{h^2 - a^2}).$$

But $$h - \sqrt{h^2 - a^2} = \frac{(h - \sqrt{h^2 - a^2})(h + \sqrt{h^2 - a^2})}{h + \sqrt{h^2 - a^2}} = \frac{a^2}{h + \sqrt{h^2 - a^2}},$$

which approaches 0 when h increases without limit.

Therefore $PQ \to 0$, and AB, or $y = bx/a$, is an asymptote to the hyperbola $b^2x^2 - a^2y^2 = a^2b^2$. § 172

Since the hyperbola is symmetric with respect to OX, the line $y = -bx/a$, or CD, is also an asymptote.

AG

174. Drawing the Hyperbola. The lines perpendicular to the real axis at the vertices A', A and the lines perpendicular to the conjugate axis at the points B', B form a rectangle. The diagonals of this rectangle are obviously the asymptotes

$$y = \frac{b}{a}x \quad \text{and} \quad y = -\frac{b}{a}x.$$

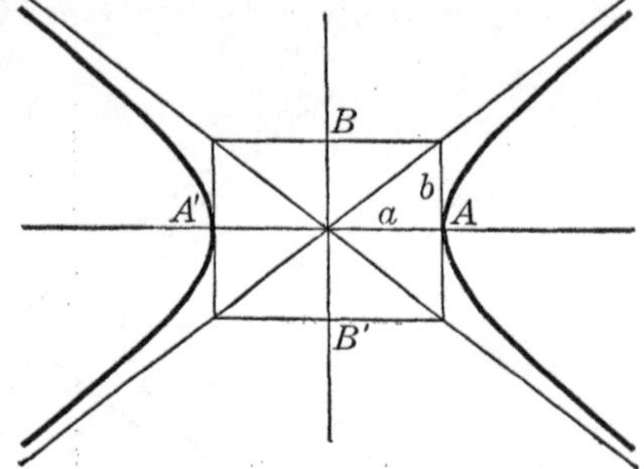

Hence, when the equation of the hyperbola

$$\frac{x^2}{a^2} - \frac{y^2}{b^2} = 1$$

is given, a simple way to draw the hyperbola is to draw this rectangle, produce its diagonals, and sketch the curve approaching these lines as asymptotes.

175. Conjugate Hyperbolas. The two hyperbolas

$$\frac{x^2}{a^2} - \frac{y^2}{b^2} = 1 \quad \text{and} \quad \frac{y^2}{b^2} - \frac{x^2}{a^2} = 1$$

are closely related. The real and conjugate axes of one are respectively the conjugate and real axes of the other. Moreover, the rectangle described in § 174 is the same for both, and therefore the two hyperbolas have the same asymptotes. The two hyperbolas are said to be *conjugate.*

176. Rectangular Hyperbola. If the hyperbola has its axes equal to each other, so that $a = b$, the equation becomes $\frac{x^2}{a^2} - \frac{y^2}{a^2} = 1$, or $x^2 - y^2 = a^2$. Hence the asymptotes are $y = x$ and $y = -x$, which are at right angles to each other. Such a hyperbola is called a *rectangular hyperbola.* It is also called an *equilateral hyperbola.*

Evidently the hyperbola $y^2 - x^2 = a^2$ is also rectangular.

Exercise 49. Equations of Hyperbolas

Draw the following hyperbolas, and find the foci, eccentricity, focal width, and directrices of each:

1. $4x^2 - 25y^2 = 100$.

2. $9x^2 - 4y^2 = 36$.

3. $4y^2 - 9x^2 = 36$.

4. $x^2 - y^2 = 64$.

5. $x^2 - y^2 = -64$.

6. $3x^2 - 8y^2 = 48$.

7. $x^2 - 4y^2 = -4$.

8. $7x^2 - 2y^2 = -63$.

9. Show that the equation $Ax^2 + By^2 = C$, where A and B have unlike signs, represents a hyperbola, and find the foci, eccentricity, and asymptotes.

Find the equations of the hyperbolas which have their axes along the coordinate axes and satisfy the following conditions, drawing the figure in each case:

10. One vertex is (4, 0), and one focus is (5, 0).

11. One vertex is (0, 8), and the eccentricity is 2.

12. One asymptote is $2y = 3x$, and one focus is (13, 0).

13. The point $(4, \sqrt{3})$ is on the curve, and (2, 0) is one vertex.

14. The points (4, 6) and (1, 1) are on the curve.

15. One asymptote is $3x - 4y = 0$, and one vertex is (0, 10).

16. The hyperbola is rectangular, and one focus is (8, 0).

17. One vertex bisects the distance from center to focus, and the focal width is 18.

18. The focal width is equal to the real axis, and one directrix is $x = 4$.

19. One focus is $F(6, 0)$, and $FP = 5$, where P is a point on the curve which has the abscissa 4.

20. If a hyperbola is rectangular, show that the eccentricity is $\sqrt{2}$ and that the focal width is equal to each of the axes.

21. Show that the four foci of two conjugate hyperbolas are equidistant from the center.

22. Every line parallel to an asymptote of a hyperbola cuts the hyperbola in one point at a finite distance from the center and in one point at an infinite distance from the center.

23. The intersections of the hyperbola $b^2x^2 - a^2y^2 = a^2b^2$ with a line through the center, which has the slope m such that $-b/a > m > b/a$, are imaginary.

24. The equation $x^2/a^2 - y^2/b^2 = 0$ represents both asymptotes of the hyperbola $x^2/a^2 - y^2/b^2 = 1$.

25. The line from a vertex of a hyperbola to one end of the conjugate axis is equal to the distance from the center to a focus.

26. If e and e' are the eccentricities of two conjugate hyperbolas, then $1/e^2 + 1/e'^2 = 1$.

27. If two hyperbolas have the same foci, show that the one that has the greater eccentricity has its vertices nearer the center. Compare the positions of the asymptotes.

28. If two hyperbolas have the same asymptotes and lie in the same pair of vertical angles of the asymptotes, they have the same eccentricity.

29. If 2θ denotes the angle between the asymptotes of a hyperbola, show that

$$\tan 2\theta = \frac{2\sqrt{e^2 - 1}}{2 - e^2}.$$

From this result show that all hyperbolas having the same eccentricity have the same angle between their asymptotes.

In § 173 it was shown that $m = b/a$, and hence $\tan\theta = b/a$.

30. The perpendiculars to the real axis at the vertices of a hyperbola meet the asymptotes in four points which lie on the circle the diameter of which is the line joining the foci.

31. If two hyperbolas have the same center and the same directrices, the distances from the center to the foci are proportional to the squares of the real axes.

THEOREM. DISTANCES FROM THE FOCI TO A POINT

177. *The difference between the distances from the foci to any point on a hyperbola is constant and equal to the real axis.*

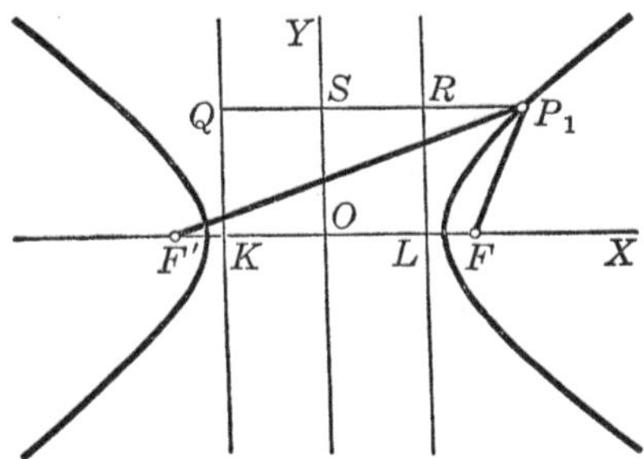

Proof. Let KQ and LR be the directrices, F' and F the corresponding foci, and e the eccentricity of the hyperbola.

If the point $P_1(x_1, y_1)$ is on the right-hand branch,

$$F'P_1 = e \cdot QP_1 = e(QS + SP_1) = e\left(\frac{a}{e} + x_1\right),$$

and $$FP_1 = e \cdot RP_1 = e(SP_1 - SR) = e\left(x_1 - \frac{a}{e}\right). \qquad \S\,164$$

That is, $$F'P_1 = ex_1 + a,$$

and $$FP_1 = ex_1 - a.$$

Therefore $$F'P_1 - FP_1 = 2a,$$

which, since $2a$ is the real axis, proves the theorem.

The above proof applies with slight changes when P_1 is on the left-hand branch of the curve, but in that case

$$FP_1 - F'P_1 = 2a.$$

We have seen that there are simple mechanical means for constructing the ellipse (§ 143) and the parabola (Ex. 20, page 118). For constructing the hyperbola there are no such simple means.

PROBLEM. HYPERBOLA WITH A GIVEN CENTER

178. *To find the equation of the hyperbola having its center at the point (h, k) and its real axis parallel to the x axis.*

Solution. Since the solution is similar to that given for the ellipse (§ 145), it is left to the student, who should find that the equation is

$$\frac{(x-h)^2}{a^2}-\frac{(y-k)^2}{b^2}=1.$$

If the real axis is parallel to the y axis, b denotes the real semiaxis of the hyperbola, and the equation is

$$\frac{(y-k)^2}{b^2}-\frac{(x-h)^2}{a^2}=1. \qquad \text{§ 171}$$

THEOREM. THE EQUATION $Ax^2 - By^2 + Cx + Dy + E = 0$

179. *If A and B have the same sign and are not 0, every equation of the form $Ax^2 - By^2 + Cx + Dy + E = 0$ represents a hyperbola having its axes parallel to the coordinate axes.*

Proof. By the process of completing squares the equation may be written $A(x-h)^2 - B(y-k)^2 = F$, where h, k, and F are constants, and this equation may then be written

$$\frac{(x-h)^2}{F/A}-\frac{(y-k)^2}{F/B}=1.$$

This equation represents a hyperbola having its axes parallel to the coordinate axes. § 178

For a more detailed explanation of the method see § 146.

In the special case when $F = 0$, the left-hand member of the equation $A(x-h)^2 - B(y-k)^2 = 0$ can be factored into two linear expressions. The equation then represents two straight lines, which we call a *degenerate hyperbola.*

Exercise 50. Drawing Hyperbolas

Draw the following hyperbolas and find the eccentricity, foci, and vertices of each:

1. $4x^2-9y^2-16x+18y=29$.

2. $9x^2-y^2+36x+6y+18=0$.

3. $3x^2-2y^2-18x-8y+1=0$.

4. $2x^2-5y^2-20x+18=0$.

5. $y^2-2x^2-12x=34$.

6. $4y^2-x^2+2x+16y=1$.

7. $x^2-y^2-8y=48$.

8. $x^2-y^2=8y$.

9. The general equation of the hyperbola having its axes parallel to OX and OY involves four essential constants.

10. The equation of a hyperbola having its axes parallel to OX and OY may be written in either of the following forms:

$$lx^2-my^2+px+qy=1, \quad (1)$$

$$l(x-h)^2-m(y-k)^2=1. \quad (2)$$

11. Find the equation of the hyperbola $4x^2-y^2=16$ when the origin is moved to the left-hand focus.

Find the equations of the hyperbolas satisfying the following conditions, and draw each figure:

12. The foci are (4, 0) and (10, 0), and one vertex is (6, 0).

If the equation is assumed in the form given in § 178, h, k, and a may be found by observation, and b is found from the fact that the distance from the center to a focus is $\sqrt{a^2+b^2}$.

13. The foci are (3, 5) and (13, 5), and the eccentricity is $\frac{5}{4}$.

14. The two vertices are $(-1, -6)$ and $(-1, 8)$, and the eccentricity is $\sqrt{2}$.

15. The two directrices are $x=-2$ and $x=4$, and one focus is $F(\frac{16}{3}, 6)$.

16. The center is (4, 1), the eccentricity is $\frac{1}{2}\sqrt{13}$, and the point (8, 4) is on the curve.

17. The hyperbola passes through the four points (5, 1), $(-3, -1)$, $(-3, 1)$, and $(-2, \frac{1}{2}\sqrt{2})$.

180. Ellipse and Hyperbola. The equation of the ellipse

$$\frac{x^2}{a^2}+\frac{y^2}{b^2}=1 \tag{1}$$

differs only in the sign of b^2 from that of the hyperbola

$$\frac{x^2}{a^2}-\frac{y^2}{b^2}=1, \quad \text{or} \quad \frac{x^2}{a^2}+\frac{y^2}{-b^2}=1. \tag{2}$$

Therefore, when certain operations performed in connection with equation (1) produce a certain result for the ellipse, the same operations performed in connection with equation (2) produce for the hyperbola a result which differs from the result for the ellipse only in the sign of b^2.

While we shall employ this method for obtaining certain results for the hyperbola, the student may obtain the same results independently by methods similar to those employed for the ellipse.

PROBLEM. SLOPE OF THE HYPERBOLA

181. *To find the slope of the hyperbola $x^2/a^2-y^2/b^2=1$ at any point $P_1(x_1, y_1)$ on the hyperbola.*

Solution. The slope of the ellipse $x^2/a^2+y^2/b^2=1$ at any point $P_1(x_1, y_1)$ on the ellipse is $-b^2x_1/a^2y_1$ (§ 147).

Then the slope m of the tangent to the hyperbola, and hence the slope of the hyperbola (§ 125), at $P_1(x_1, y_1)$ is

$$m=\frac{b^2x_1}{a^2y_1}. \qquad \text{§ 180}$$

182. COROLLARY. *The slope m' of the conjugate hyperbola $y^2/b^2-x^2/a^2=1$ at any point $P_1(x_1, y_1)$ on the hyperbola is*

$$m'=\frac{b^2x_1}{a^2y_1}.$$

The proof is left to the student. Notice that the equations of the conjugate hyperbola and ellipse differ only in the sign of a^2.

PROBLEM. TANGENT TO THE HYPERBOLA

183. *To find the equation of the tangent to the hyperbola $x^2/a^2 - y^2/b^2 = 1$ at the point $P_1(x_1, y_1)$ on the hyperbola.*

Solution. Applying § 180 to the result of § 148, we have

$$\frac{x_1 x}{a^2} - \frac{y_1 y}{b^2} = 1.$$

184. COROLLARY 1. *The equation of the normal to the hyperbola $x^2/a^2 - y^2/b^2 = 1$ at the point $P_1(x_1, y_1)$ is*

$$y - y_1 = -\frac{a^2 y_1}{b^2 x_1}(x - x_1).$$

The proof of §§ 184 and 185 is left to the student.

185. COROLLARY 2. *The intercepts of the tangent and the normal to the hyperbola $x^2/a^2 - y^2/b^2 = 1$ at the point $P_1(x_1, y_1)$ are as follows:*

1. x intercept of tangent, $x = \frac{a}{x_1}$;
2. y intercept of tangent, $y = -\frac{b^2}{y_1}$;
3. x intercept of normal, $x = \frac{a^2 + b^2}{a^2} x_1 = e^2 x_1$;
4. y intercept of normal, $y = \frac{a^2 + b^2}{b^2} y_1$.

PROBLEM. TANGENTS HAVING A GIVEN SLOPE

186. *To find the equations of the lines which have the slope m and are tangent to the hyperbola $x^2/a^2 - y^2/b^2 = 1$.*

Solution. Applying § 180 to the result of § 151, we have

$$y = mx \pm \sqrt{a^2 m^2 - b^2}.$$

Hence there are two tangents which have the slope m.

Theorem. Tangent to the Hyperbola

187. *The tangent to a hyperbola at any point* $P_1(x_1, y_1)$ *on the hyperbola bisects the angle between the focal radii of the point.*

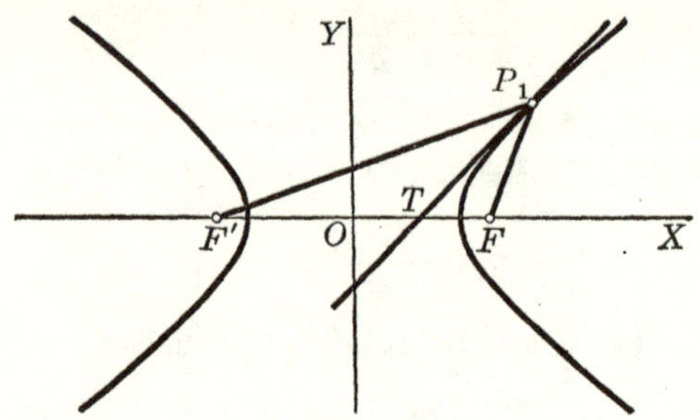

Proof. Since $F'O = OF$, § 167

$OF = ae$, § 164

and $OT = \frac{a^2}{x_1}$, § 185

we have $$\frac{F'T}{TF} = \frac{F'O + OT}{OF - OT} = \frac{ae + \frac{a^2}{x_1}}{ae - \frac{a^2}{x_1}} = \frac{ex_1 + a}{ex_1 - a}.$$

Since $F'P_1 = ex_1 + a$

and $FP_1 = ex_1 - a$, § 177

we have $$\frac{F'T}{TF} = \frac{F'P_1}{FP_1}.$$

Thus, P_1T divides $F'F$ into segments proportional to $F'P_1$ and FP_1, and therefore P_1T bisects the angle $F'P_1F$.

188. Corollary. *The tangent at* P_1 *bisects two vertical angles formed by producing the focal radii, and the normal at* P_1 *bisects the other two vertical angles so formed.*

Exercise 51. Tangents and Normals

In each of the following cases show that P is on the hyperbola, find the equations of the tangent and the normal at P, and draw the figure:

1. $4x^2 - y^2 = 64$; $P(5, 6)$.

2. $x^2 - 9y^2 = 25$; $P(-13, 4)$.

3. $y^2 - x^2 = 16$; $P(3, 5)$.

4. $2x^2 - 3y^2 = 50$; $P(7, -4)$.

5. Find the equations of the lines which are tangent to the hyperbola $16x^2 - 25y^2 = 400$ and parallel to the line $2x - 2y = 5$.

6. Draw the hyperbola $x^2 - y^2 = 16$, the tangent at the point $P(5, 3)$, and another tangent perpendicular to OP. Find the equations of both tangents.

7. Find the points of contact of the lines which are tangent to the hyperbola $4x^2 - 3y^2 = 96$ and parallel to the line $y = 2x$.

8. Find h and k such that the tangent to the hyperbola $5x^2 - 2y^2 = 18$ at the point (h, k) passes through $(1, -4)$.

9. Find m such that the line with the slope m and tangent to the hyperbola $x^2 - y^2 = 9$ passes through $(3, 9)$.

10. Prove that from any point in the plane two tangents can be drawn to a hyperbola. Under what conditions are these tangents real and distinct, real and coincident, or imaginary?

11. Find the equations of the lines passing through the point $(-6, -1)$ and tangent to the hyperbola $9x^2 - 25y^2 = 225$.

12. If the normal to the hyperbola $x^2 - y^2 = 7$ at the point (h, k) on the curve passes through $(0, 6)$, find h and k.

13. Find the point of contact of that tangent to the hyperbola $x^2 - 4y^2 = 16$ which has equal intercepts on the axes.

14. If a tangent is drawn to a rectangular hyperbola, the subnormal is equal to the abscissa of the point of contact.

15. Find each point on the hyperbola $9x^2 - 25y^2 = 225$ for which the subtangent is equal to the subnormal.

16. Show how to construct a tangent and a normal to a hyperbola at any point on the curve.

A simple construction may be found by the results of § 185.

17. Find the condition under which the line $x/h + y/k = 1$ is tangent to the hyperbola $x^2/a^2 - y^2/b^2 = 1$.

If the line is tangent, its equation may be written in another form, $x_1x/a^2 - y_1y/b^2 = 1$. This being done, we may compare the two equations of the line (§ 89) and use the fact that (x_1, y_1) is on the hyperbola.

18. Find the equation of a tangent drawn to the hyperbola $x^2 - y^2 = 4$ and having a segment $\sqrt{15}$ between the axes.

19. There is no tangent to the hyperbola $b^2x^2 - a^2y^2 = a^2b^2$ which has a slope less than b/a and greater than $-b/a$.

20. If $b > a$, no two real tangents to the hyperbola $b^2x^2 - a^2y^2 = a^2b^2$ can be perpendicular to each other.

21. Find the locus of the foot of the perpendicular from a focus of a hyperbola to a variable tangent.

See note to Ex. 14, page 155.

22. Prove that the locus of the intersection of two perpendicular tangents to a hyperbola is a circle. State the conditions under which this circle is real or imaginary.

Compare Ex. 20, above.

23. Find the locus of the foot of the perpendicular from the center of a hyperbola to a variable tangent.

24. No tangent to a hyperbola is tangent to the conjugate hyperbola.

25. All the tangents at points having the same abscissa to two or more hyperbolas having the same vertices meet on the real axis.

26. If the tangent to the hyperbola $2x^2 - ky^2 = l$ at (4, 2) passes through (1, − 2), find k and l.

27. If a hyperbola is rectangular, find the equation of a tangent passing through the focus of the conjugate hyperbola.

In this figure given that P_1T is the tangent to the hyperbola at $P_1(x_1, y_1)$, P_1N is the normal, SP_1 is perpendicular to OX, and FE, OQ, and $F'E'$ are perpendicular to P_1T, establish the following properties of the hyperbola:

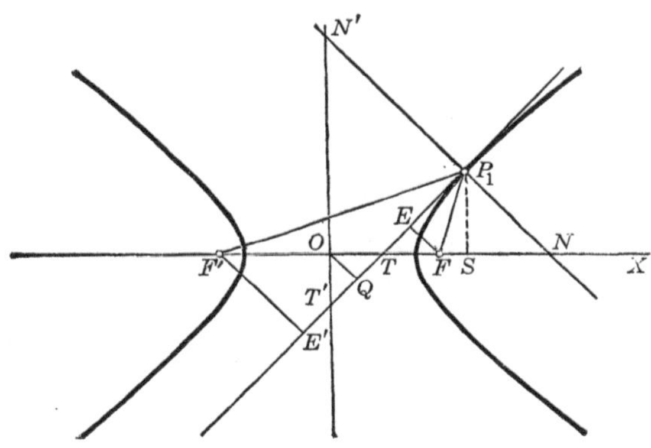

28. $TS = \dfrac{x_1^2 - a^2}{x_1}$.

29. $ON \cdot OT = \overline{OF}^2$.

30. $OS \cdot OT = a^2$.

31. $OQ \cdot NP_1 = -b^2$.

32. $OQ \cdot N'P_1 = a^2$.

33. $NS = -\dfrac{b^2}{a^2}x_1$.

34. $N'P_1 \cdot P_1N = F'P_1 \cdot FP_1$.

35. $N'P_1 \cdot P_1N = T'P_1 \cdot TP_1$.

36. $F'E' \cdot FE = -b^2$.

37. $TF \cdot OS = a \cdot FP_1$.

The student will find it interesting to compare these properties of the hyperbola with the similar list for the ellipse on page 155. In Ex. 32 draw P_1S' perpendicular to OY and observe that the triangles $T'TO$ and $T'P_1N'$ are similar. Then $OQ : S'P_1 = OT : N'P_1$, whence $OQ \cdot N'P_1 = S'P_1 \cdot OT$.

38. If a variable tangent to a hyperbola cuts the asymptotes at the points A and B, then $OA \cdot OB$ is constant.

39. The product of the distances to the asymptotes to a hyperbola from a variable point on the curve is constant.

40. If the hyperbola in the above figure is rectangular, SP_1 is the mean proportional between OS and TS.

41. Only one normal to a hyperbola passes between the foci.

189. Diameter. By §§ 153 and 180 the equation of the diameter which bisects the system of chords of the hyperbola $b^2x^2 - a^2y^2 = a^2b^2$ that have the slope m is

$$y = \frac{b^2}{a^2m}x.$$

If m' is the slope of this diameter, then $m'm = b^2/a^2$.

Theorem. Conjugate Diameters

190. *If one diameter of a hyperbola bisects the chords parallel to another diameter, then the second diameter bisects the chords parallel to the first.*

Proof. The condition under which the diameter with slope m' bisects that with slope m is $m'm = b^2/a^2$ (§ 189).

But this is also the condition under which the second diameter bisects the chords parallel to the first diameter.

191. Conjugate Diameters. If each of two diameters of a hyperbola bisects the chords parallel to the other, the diameters are called *conjugate diameters.*

Theorem. Position of Conjugate Diameters

192. *In every pair of conjugate diameters of the hyperbola $b^2x^2 - a^2y^2 = a^2b^2$, the diameters pass through the same quadrant and lie on opposite sides of the asymptote in that quadrant.*

Proof. Since $m'm = b^2/a^2$, the slopes of two conjugate diameters are both positive or both negative. Hence the diameters pass through the same quadrant.

Since $m'm = \frac{b^2}{a^2}$, if $|m| < \frac{b}{a}$ we see that $|m'| > \frac{b}{a}$.

By $|m|$ is meant the numerical value of m, irrespective of sign.

But the slopes of the asymptotes are b/a and $-b/a$ (§ 173), and so the conjugate diameters lie as stated.

THEOREM. ENDS OF CONJUGATE DIAMETERS

193. *Of two conjugate diameters one meets the hyperbola in real points and the second does not; but the second diameter meets the conjugate hyperbola in real points.*

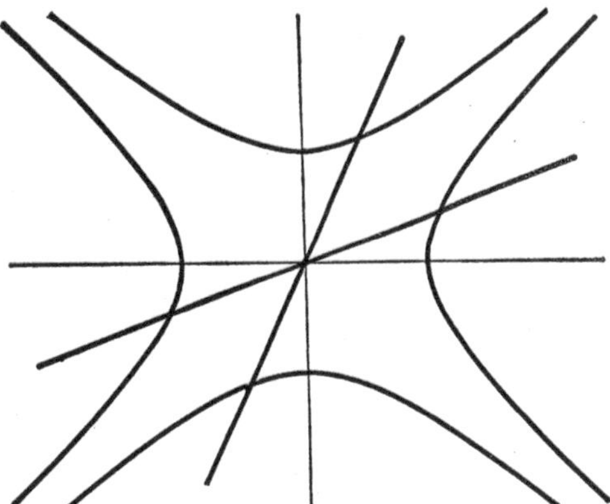

Proof. Let $y = mx$ and $y = m'x$ be the conjugate diameters, and take $|m| < b/a$ and $|m'| > b/a$. § 192

The abscissas of the points in which the diameter $y = mx$ meets the hyperbola $b^2x^2 - a^2y^2 = a^2b^2$ are the two roots of the quadratic equation $b^2x^2 - a^2(mx)^2 = a^2b^2$.

Hence $$x = \pm \frac{ab}{\sqrt{b^2 - a^2m^2}}.$$

But since $|m| < b/a$, it follows that $a^2m^2 < b^2$, and hence these roots are real.

It is left to the student to prove that the diameter $y = m'x$ does not meet the hyperbola $b^2x^2 - a^2y^2 = a^2b^2$ in real points, but does meet the conjugate hyperbola $a^2y^2 - b^2x^2 = a^2b^2$ (§ 175) in real points.

194. Ends of the Diameters. The real points in which the conjugate diameters meet the two hyperbolas are called the *ends of the diameters.*

THEOREM. DIAMETER OF THE CONJUGATE HYPERBOLA

195. *The diameter bisecting those chords of the hyperbola* $\frac{x^2}{a^2}-\frac{y^2}{b^2}=1$ *which have the slope m bisects also those chords of the conjugate hyperbola* $\frac{y^2}{b^2}-\frac{x^2}{a^2}=1$ *which have the slope m.*

Proof. The equations of the hyperbolas differ only in the signs of a^2 and b^2. But the equation of the diameter bisecting those chords of the hyperbola $\frac{x^2}{a^2}-\frac{y^2}{b^2}=1$ which have the slope m is $y=\frac{b^2}{a^2m}x$ (§ 189). Therefore this is also the equation of the diameter bisecting those chords of the conjugate hyperbola which have the slope m.

THEOREM. CONJUGATE DIAMETERS

196. *Two diameters which are conjugate with respect to a hyperbola are also conjugate with respect to the conjugate hyperbola.*

Proof. The condition under which two diameters are conjugate with respect to the one hyperbola is $m'm=\frac{b^2}{a^2}$ (§ 190). By the method of § 195 it is also the condition under which they are conjugate with respect to the conjugate hyperbola.

PROBLEM. ENDS OF A CONJUGATE DIAMETER

197. *Given an end* $P(x_1, y_1)$ *of a diameter of the hyperbola* $b^2x^2-a^2y^2=a^2b^2$, *to find the ends of the conjugate diameter.*

It is left to the student to show, by the method employed in the case of the ellipse (§ 158), that the ends of the conjugate diameter on the conjugate hyperbola are $\left(\frac{a}{b}y_1, \frac{b}{a}x_1\right)$ and $\left(-\frac{a}{b}y_1, -\frac{b}{a}x_1\right)$.

Exercise 52. Diameters

1. Draw the hyperbola $16x^2 - 9y^2 = 144$, the chord made by the line $x = y + 10$, and the diameter which bisects this chord. Find the equation of the diameter.

2. Draw the diameter conjugate to that of Ex. 1, and find its equation.

3. Draw the hyperbola $9x^2 - 4y^2 = 36$ and the chord having the mid point (4, 3). Find the equation of the chord.

4. Find the mid point of the chord determined by the hyperbola $4x^2 - 25y^2 = 100$ and the line $5y = x + 15$.

5. Draw the hyperbola $y^2 - x^2 = 4$, and find the equation of the chord through the point $(-2, -2)$ and bisected by the diameter having the slope $\frac{1}{2}$.

6. Find the equations of two conjugate diameters of the hyperbola $3x^2 - 4y^2 = 48$, given that one meets the curve at the point having the abscissa 8 and a negative ordinate.

7. The tangents at the ends of a diameter of a hyperbola are parallel to each other and are also parallel to the conjugate diameter.

8. Draw two conjugate hyperbolas and also draw a straight line cutting them in four real points. Denote these points in order by A, B, C, D, and prove that $AB = CD$.

9. If the tangent to a hyperbola at P meets the conjugate hyperbola at A and B, then $AP = PB$.

10. If a' and b' are conjugate semidiameters of a hyperbola, then $a'^2 - b'^2$ is constant.

11. If θ is the angle from a diameter to its conjugate, then $\sin\theta = ab/a'b'$.

Compare the corresponding theorem for the ellipse (§ 159).

12. The area of the parallelogram formed by the tangents to two conjugate hyperbolas at the ends of conjugate diameters is constant.

Problem. Asymptotes as Axes

198. *To find the equation of a hyperbola referred to the asymptotes as axes of coordinates.*

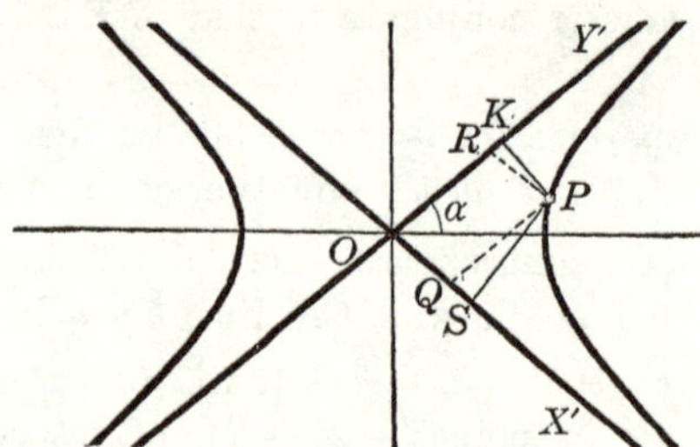

Solution. The equations of the asymptotes OY' and OX' of the hyperbola $b^2x^2 - a^2y^2 = a^2b^2$ are (§ 173) $bx - ay = 0$ and $bx + ay = 0$. Therefore, if PS and PK are the perpendiculars from $P(x, y)$ to OX' and OY' respectively, by § 84

$$SP \cdot KP = \frac{bx + ay}{\sqrt{a^2 + b^2}} \cdot \frac{bx - ay}{\sqrt{a^2 + b^2}} = \frac{b^2x^2 - a^2y^2}{a^2 + b^2} = \frac{a^2b^2}{a^2 + b^2}. \quad (1)$$

Denote by 2α the angle $X'OY'$, and by x' and y' the coordinates of P when OX' and OY' are taken as axes. Then $x' = RP$ and $y' = QP$. Then $SP = y' \sin 2\alpha$, $KP = x' \sin 2\alpha$, and hence

$$x'y' \sin^2 2\alpha = \frac{a^2b^2}{a^2 + b^2}. \quad (2)$$

But $\tan\alpha$, which is the slope of OY', is $\frac{b}{a}$. Therefore

$$\sin\alpha = \frac{b}{\sqrt{a^2 + b^2}}, \quad \cos\alpha = \frac{a}{\sqrt{a^2 + b^2}},$$

and

$$\sin 2\alpha = 2 \sin\alpha \cos\alpha = \frac{2ab}{a^2 + b^2}.$$

Hence, dropping the primes in (2), we have

$$\mathbf{4\,xy = a^2 + b^2.}$$

199. Rectangular Hyperbola referred to Asymptotes. The result of § 198 is of special interest in the case of the rectangular hyperbola, for which the asymptotes are perpendicular to each other. In this case, $a = b$, and the equation

$$4\,xy = a^2 + b^2$$

may be written

$$2\,xy = a^2,$$

or $$xy = \tfrac{1}{2}\,a^2.$$

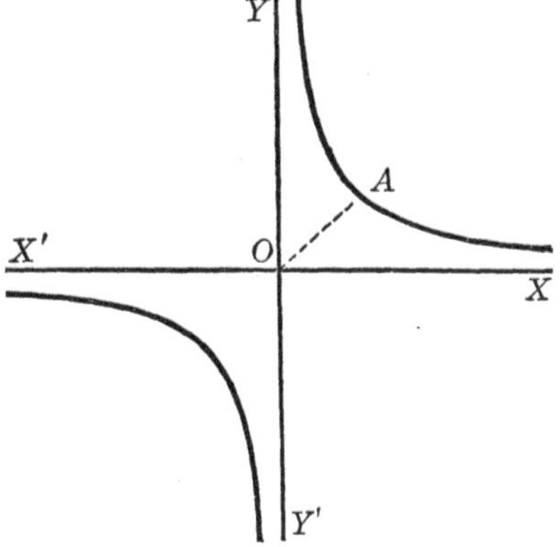

This figure shows a rectangular hyperbola drawn with its asymptotes in the horizontal and vertical position.

Therefore the equation $xy = c$, where c is any constant excepting 0, is the equation of a hyperbola referred to its asymptotes as axes. If c is negative, then the values of x and y must be unlike in sign, and hence the hyperbola lies in the second and fourth quadrants.

In the case of the rectangular hyperbola, the equation of which is $xy = c$, where c is positive, the semiaxis a, or OA, may be found from the fact that $c = \frac{1}{2}\,a^2$. Thus, for the distance OA from the center to the vertex of the rectangular hyperbola $xy = 18$, we have $18 = \frac{1}{2}\,a^2$; whence $a = 6$.

Boyle's Law of Gases, $vp = c$, which states that the product of the volume of a gas and the pressure upon the gas is constant, is represented graphically by the upper branch of the above figure, since there are no negative values of v and p.

There are many other laws of physics and chemistry which are expressed by the equation $xy = k$, a constant. Similarly, in mensuration we frequently have cases in which $xy = k$, as for example in the formula for the area of a rectangle, the area of an ellipse, and so on.

Exercise 53. Review

1. The slope of a focal chord of the hyperbola $x^2 - y^2 = 8$ is 7. In what ratio is the chord divided by the focus?

2. Given the hyperbola $x^2 - 4y^2 = 100$, find the slope of a chord which passes through the center and has the length 30.

3. Show that the equation $y = \pm\sqrt{4x^2 - 8x}$ represents a hyperbola, and draw the curve.

4. Consider Ex. 3 for the equation $y = \pm\sqrt{4x^2 + 8x}$.

5. If the asymptotes of a hyperbola are perpendicular to each other, the semiaxes a and b are equal.

6. Given that a line through a focus F of a rectangular hyperbola and parallel to an asymptote cuts the curve at G, find the length of FG.

7. Given that a perpendicular to the real axis of a hyperbola at M meets the curve at P and an asymptote at Q, prove that $\overline{MQ}^2 - \overline{MP}^2$ is constant.

8. The distance from the center of a rectangular hyperbola to any point P on the curve is the mean proportional between the distances from the foci to P.

9. All hyperbolas having the same foci and the same asymptotes coincide.

10. If two hyperbolas have the same asymptotes, they have the same eccentricity or else the sum of the squares of the reciprocals of their eccentricities is 1.

11. Find the eccentricity of every hyperbola having the asymptotes $y = \pm mx$.

12. What does the equation of the hyperbola $9x^2 - 4y^2 = 36$ become when the asymptotes are taken as axes?

13. If the asymptotes are perpendicular to each other, find the coordinates of the vertices of the hyperbola $xy = 36$.

14. Consider Ex. 13 for the case in which the angle between the asymptotes is 2α.

15. Find the equation of the hyperbola having the directrix $3x - 4y = 10$, the corresponding focus (6, 0), and the eccentricity 2.

16. The distance from a focus of a hyperbola to an asymptote is equal to the conjugate semiaxis.

17. The foci of a hyperbola are F' and F, a tangent t is drawn to the curve at any point P, and the perpendicular from F upon t meets $F'P$ at H. Prove that $F'H$ is equal to the real axis of the hyperbola.

18. The segment of a tangent to a hyperbola intercepted between the tangents at the vertices subtends a right angle at each focus.

19. A focus of a hyperbola is F, a focal ordinate is FQ, a tangent t is drawn to the hyperbola at Q, and the ordinate MP of any point P on the curve is produced to meet the tangent t at R. Prove that $FP = MR$.

20. If the circle through the foci of a hyperbola and any point P on the curve cuts the conjugate axis at Q and R, then the tangent at P passes through one of the points Q, R, and the normal at P passes through the other.

The analytic proof is laborious. The use of § 187 leads to an easy geometric proof.

21. Any two conjugate diameters of a rectangular hyperbola are equal.

22. The tangents at any two points P_1 and P_2 on a hyperbola meet on the diameter which bisects P_1P_2.

23. If the line through the points Q and R on a hyperbola meets the asymptotes at P and S, the mid points of PS and QR coincide, and hence $PQ = RS$.

24. If the ordinate drawn from any point P on the hyperbola $b^2x^2 - a^2y^2 = a^2b^2$ is produced, if necessary, to meet the rectangular hyperbola $x^2 - y^2 = a^2$ at Q, the ratio of the ordinates of P and Q is constant.

Given the system of hyperbolas $x^2/a^2 - y^2/b^2 = k$, where k takes various values but a and b are fixed, prove that:

25. Through each point of the plane there passes one and only one hyperbola of the system.

26. All the hyperbolas have the same asymptotes.

27. All the hyperbolas for which k is positive have the same eccentricity, and so do all those for which k is negative.

Letting E represent the equation $\dfrac{x^2}{a^2+k} + \dfrac{y^2}{b^2+k} = 1$, prove the statements in Exs. 28–31:

28. For each negative value of k numerically between a^2 and b^2, E represents a hyperbola; for each other value of k, E represents an ellipse.

29. All the curves E have the same foci.

30. Given that $a^2 = 4$ and $b^2 = 1$, through $(2, 1)$ two of the conics E, one an ellipse and one a hyperbola, can be passed.

31. The two conics E of Ex. 30 intersect at right angles.

A simple proof is suggested by §§ 152, 187, and 188.

32. Given a hyperbola, show how to find by construction the center and axes.

33. The locus of the center of a circle which is tangent to two fixed circles is a hyperbola the foci of which are the centers of the fixed circles.

34. Find the locus of the center of a circle which cuts from the axes chords of constant length $2a$ and $2b$.

35. Given that two vertices of a triangle are $A(3, 0)$ and $B(-3, 0)$, find the locus of the third vertex P if P moves so that the angle ABP is twice the angle PAB, draw the locus, and show how to trisect any angle by means of it.

36. If two hyperbolas have the same vertices, and the perpendicular to the real axis at any point M cuts one hyperbola at A and the other at B, then MA and MB are to each other as the conjugate axes of the respective hyperbolas.

CHAPTER X

CONICS IN GENERAL

THEOREM. EQUATION OF THE SECOND DEGREE

200. *Every equation of the second degree in rectangular coordinates represents a conic.*

Proof. The general equation in x and y is

$$ax^2 + 2\,hxy + by^2 + 2\,gx + 2\,fy + c = 0,$$

where a, h, b, g, f, and c are arbitrary constants, but a, h, and b are not all three equal to 0.

By rotating the axes through a properly chosen angle (§ 111) we can eliminate the xy term, the equation becoming

$$a'x^2 + b'y^2 + 2\,g'x + 2\,f'y + c' = 0,$$

an equation which represents an ellipse (§ 146), a hyperbola (§ 179), or a parabola (§§ 122, 123).

THEOREM. EQUATION OF ANY CONIC

201. *The equation of any conic referred to rectangular coordinates is of the second degree.*

Proof. The equation of any conic referred to axes taken through any origin O parallel to certain lines is

$$Ax^2 + By^2 + Cx + Dy + E = 0.$$

The equation of the same conic referred to axes through O making any angle θ with the other axes is found by substitutions (§ 109) which do not change the degree.

Problem. Classification of the Conics

202. *To find the conditions under which the general equation* $ax^2 + 2hxy + by^2 + 2gx + 2fy + c = 0$ *represents a parabola, an ellipse, or a hyperbola.*

Solution. Rotate the axes through an angle θ such that $\tan 2\theta = 2h/(a-b)$, thus eliminating the xy term (§ 111). Putting $x \cos\theta - y \sin\theta$ for x and $x \sin\theta + y \cos\theta$ for y, the general equation takes the form

$$a'x^2 + b'y^2 + (\text{terms of lower degree}) = 0, \tag{1}$$

where
$$a' = a\cos^2\theta + b\sin^2\theta + 2h\sin\theta\cos\theta, \tag{2}$$

and
$$b' = a\sin^2\theta + b\cos^2\theta - 2h\sin\theta\cos\theta. \tag{3}$$

We know that (1) represents a parabola if either $a' = 0$ or $b' = 0$ (§§ 122, 123); an ellipse if a' and b' have like signs (§ 146); or a hyperbola if a' and b' have unlike signs (§ 179). Furthermore, we may express these three conditions in terms of a, b, and h by means of (2), (3), and the fact that

$$\tan 2\theta = 2h/(a-b) = \sin 2\theta/\cos 2\theta;$$

that is,
$$0 = (a-b)\sin 2\theta - 2h\cos 2\theta. \tag{4}$$

Adding (2) and (3), then subtracting (3) from (2) and recalling that $\sin^2\theta + \cos^2\theta = 1$, $\cos^2\theta - \sin^2\theta = \cos 2\theta$, and $2\sin\theta\cos\theta = \sin 2\theta$, we have

$$a' + b' = a + b, \tag{5}$$

and
$$a' - b' = (a-b)\cos 2\theta + 2h\sin 2\theta. \tag{6}$$

Eliminating θ by squaring (4) and (6) and adding, we have

$$(a'-b')^2 = (a-b)^2 + 4h^2. \tag{7}$$

Subtracting (7) from the square of (5), we have

$$\boldsymbol{a'b' = ab - h^2.}$$

We may now make the following summary:

If $ab - h^2 = 0$, the given conic is a parabola.

For then either $a' = 0$ or $b' = 0$.

If $ab - h^2 > 0$, the conic is an ellipse.

For then a' and b' must have like signs.

If $ab - h^2 < 0$, the conic is a hyperbola.

For then a' and b' must have unlike signs.

Since $ax^2 + 2\,hx + b$ has equal factors if $(2\,h)^2 - 4\,ab = 0$, we see that $ax^2 + 2\,hxy + by^2$ is a perfect square if $h^2 - ab = 0$. Then the conic is a parabola when the terms of the second degree form a perfect square, as in the equation $4\,x^2 + 12\,xy + 9\,y^2 - 3\,x + y = 7$.

203. Central Conics. The ellipse and hyperbola have centers and hence are called *central conics*. If $ab - h^2 \neq 0$, an equation of the second degree represents a central conic.

204. Standard Equations. We shall refer to the following as the *standard equations* of the conics:

For the parabola, $y^2 = 4\,px.$ § 115

For the ellipse, $\dfrac{x^2}{a^2} + \dfrac{y^2}{b^2} = 1.$ § 136

For the hyperbola, $\dfrac{x^2}{a^2} - \dfrac{y^2}{b^2} = \pm 1.$ § 165

We now proceed to the problem of reducing the general equation of the second degree to these forms.

For the central conics, we move the origin in order to eliminate the x terms and y terms (§ 107), and we then rotate the axes in order to eliminate the xy term (§ 111).

205. Choosing θ. There are always two values of $2\,\theta$ less than 360° for which $\tan 2\,\theta = 2\,h/(a-b)$, and one of them is less than 180°, in which case $\theta < 90°$. In the treatment of central conics we shall always take $\theta < 90°$.

Problem. Moving the Origin for Central Conics

206. *Given the equation of a central conic referred to any axes,* $ax^2 + 2hxy + by^2 + 2gx + 2fy + c = 0$, *where* $ab - h^2 \neq 0$, *to move the origin to the center.*

Solution. To move the origin to any point $O'(m, n)$ we must write $x + m$ for x and $y + n$ for y in the given equation (§ 105). This gives the equation

$$ax^2 + 2hxy + by^2 + 2(am + hn + g)x$$
$$+ 2(hm + bn + f)y + c' = 0, \qquad (1)$$

where $c' = am^2 + 2hmn + bn^2 + 2gm + 2fn + c$.

Now choose m and n so that the coefficients of the x term and the y term are 0; that is, so that

$$am + hn + g = 0,$$

and

$$hm + bn + f = 0; \qquad (2)$$

or so that

$$m = \frac{hf - bg}{ab - h^2},$$

and

$$n = \frac{hg - af}{ab - h^2}. \qquad (3)$$

Then the transformed equation becomes

$$\mathbf{ax^2 + 2hxy + by^2 + c' = 0.} \qquad (4)$$

The test for symmetry (§ 43) shows that the graph of this equation is symmetric with respect to the new origin O', which is therefore the center of the conic. Hence equation (4) is the required equation.

207. Corollary. *The coordinates* (m, n) *of the center of a central conic* $ax^2 + 2hxy + by^2 + 2gx + 2fy + c = 0$ *are*

$$m = \frac{hf - bg}{ab - h^2} \quad \textit{and} \quad n = \frac{hg - af}{ab - h^2}.$$

PROBLEM. TO ELIMINATE THE xy TERM

208. *To transform the equation* $ax^2 + 2hxy + by^2 + c' = 0$ *of a central conic into an equation which has no* xy *term.*

Solution. Rotate the axes through an angle θ such that $\tan 2\theta = 2h/(a-b)$ (§ 111), writing $x\cos\theta - y\sin\theta$ for x and $x\sin\theta + y\cos\theta$ for y. This gives $a'x^2 + b'y^2 + c' = 0$,

or
$$\frac{x^2}{-\frac{c'}{a'}} + \frac{y^2}{-\frac{c'}{b'}} = 1. \qquad (1)$$

This completes the reduction of any equation of a central conic to the standard form. It is desirable, however, to have formulas giving the values of a', b', and c' in terms of the coefficients of the general equation.

By § 202, $a' + b' = a + b$ and $a' - b' = \pm\sqrt{(a-b)^2 + 4h^2}$.

Therefore
$$a' = \tfrac{1}{2}\left(a + b \pm \sqrt{(a-b)^2 + 4h^2}\right) \qquad (2)$$

and
$$b' = \tfrac{1}{2}\left(a + b \mp \sqrt{(a-b)^2 + 4h^2}\right). \qquad (3)$$

Whether to use the upper or the lower signs before the radicals depends upon the value of θ. From $\tan 2\theta = 2h/(a-b)$ we obtain

$$\sin 2\theta = 2h/\sqrt{(a-b)^2 + 4h^2} = 2h/(a'-b').$$

But $2\theta < 180°$ (§ 205), and hence $\sin 2\theta$ is positive. Therefore $a' - b'$ must have the same sign as h, and a' and b' must be so chosen.

We also have (§ 206) the following relations:

$$\begin{aligned} c' &= am^2 + 2hmn + bn^2 + 2gm + 2fn + c \\ &= m(am + hn + g) + n(hm + bn + f) + gm + fn + c \\ &= gm + fn + c. && \text{§ 206, (2)} \end{aligned}$$

Hence
$$c' = \frac{abc - af^2 - bg^2 - ch^2 + 2fgh}{ab - h^2}. \qquad \text{§ 207}$$

PROBLEM. REDUCTION FOR THE PARABOLA

209. *To reduce to the standard form the equation of the parabola* $ax^2 + 2hxy + by^2 + 2gx + 2fy + c = 0$ *where it is given that* $ab - h^2 = 0$.

Solution. Since $ab - h^2 = 0$, $ax^2 + 2hxy + by^2$ is a perfect square (§ 202), and the given equation may be written

$$(sy + rx)^2 + 2gx + 2fy + c = 0, \qquad (1)$$

where $s = \sqrt{b}$ and $r = \pm\sqrt{a}$, s being always positive and r having the same sign as h. It is obvious that $rs = h$.

We now rotate the axes through an angle θ such that $\tan 2\theta = 2h/(a - b)$, that is, such that

$$\frac{2\tan\theta}{1 - \tan^2\theta} = \frac{2rs}{r^2 - s^2};$$

whence $$\tan\theta = -\frac{r}{s} \text{ or } \frac{s}{r}.$$

Any resulting value of θ serves to remove the xy term (§ 111) from the given equation. The reduction is then completed by moving the origin to the vertex.

If we choose $\tan\theta = -r/s$, we see that if r is negative, $\tan\theta$ is positive, and we take $\theta < 90°$; if r is positive, $\tan\theta$ is negative, and we take $270° < \theta < 360°$. In either case,

$$\sin\theta = -r/\sqrt{r^2 + s^2},$$

and $$\cos\theta = s/\sqrt{r^2 + s^2}.$$

In (1), putting $x\cos\theta - y\sin\theta$ for x and $x\sin\theta + y\cos\theta$ for y (§ 109), that is, putting $(sx + ry)/\sqrt{a + b}$ for x and $(-rx + sy)/\sqrt{a + b}$ for y, we have

$$Ay^2 + By + Dx + c = 0,$$

where $A = a + b$, $B = 2\dfrac{rg + sf}{\sqrt{a + b}}$, and $D = 2\dfrac{sg - rf}{\sqrt{a + b}}$.

By completing the square (§ 122), we reduce the equation $Ay^2 + By + Dx + c = 0$ to the form

$$A(y-q)^2 = -D(x-p),$$

where $q = -B/2A$, and $p = B^2/4AD - c/D$.

Then moving the origin to the vertex (p, q), we have

$$y^2 = -\frac{D}{A}x; \qquad \text{§§ 105, 121}$$

that is,
$$y^2 = 2\frac{rf - sg}{(a+b)^{\frac{3}{2}}}x.$$

PROBLEM. DEGENERATE CONICS

210. *Given the equation* $ax^2+2hxy+by^2+2gx+2fy+c=0$, *to find the condition under which it represents two straight lines.*

Solution. If the equation represents two straight lines, that is, if the conic is degenerate (§ 54), the left member can be factored (§ 179, note) thus:

$$ax^2 + 2hxy + by^2 + 2gx + 2fy + c = (lx + my + n)(px + qy + r).$$

Then
$$a = lp, \quad b = mq, \quad c = nr,$$
$$2h = lq + mp, \quad 2g = lr + np, \quad 2f = mr + nq.$$

These conditions are expressed in terms of a, b, c, f, g, h thus:

$$2f \cdot 2g \cdot 2h = 2lmnpqr + lp(m^2r^2 + n^2q^2) + mq(l^2r^2 + n^2p^2) + nr(l^2q^2 + m^2p^2);$$

that is,
$$8fgh = 2abc + a(\overline{2f}^2 - 2bc) + b(\overline{2g}^2 - 2ac) + c(\overline{2h}^2 - 2ab),$$

or
$$\boldsymbol{abc + 2fgh - af^2 - bg^2 - ch^2 = 0.}$$

211. Discriminant. The left member, that is, the expression $abc + 2fgh - af^2 - bg^2 - ch^2$, is called the *discriminant* of the equation of the second degree, and is denoted by Δ.

212. Illustrative Examples. 1. Determine the nature of the conic $34x^2+24xy+41y^2-20x+140y+50=0$ and draw the figure.

Here $ab-h^2=34\times 41-12^2>0$, and the conic is an ellipse (§ 202).

To find the center $O'(m, n)$, we have, by § 206 (2),

$$34m+12n-10=0$$

and

$$12m+41n+70=0,$$

whence $m=1$ and $n=-2$, so that the center is $O'(1, -2)$.

The equation of the conic referred to axes $O'Q$, $O'R$ through O' and parallel to OX and OY is (§ 206)

$$34x^2+24xy+41y^2+c'=0,$$

where, by § 208,

$$\begin{aligned} c' &= gm+fn+c \\ &= -10\times 1+70(-2)+50 \\ &= -100. \end{aligned}$$

Now turn the axes through the angle θ such that (§ 208)

$$\tan 2\theta=\frac{2h}{a-b}=-\frac{24}{7},$$

and construct the angle 2θ accordingly, making it less than 180°; then construct θ. The line $O'X'$ making the angle θ with OX is the new x axis, and by § 208 the transformed equation is

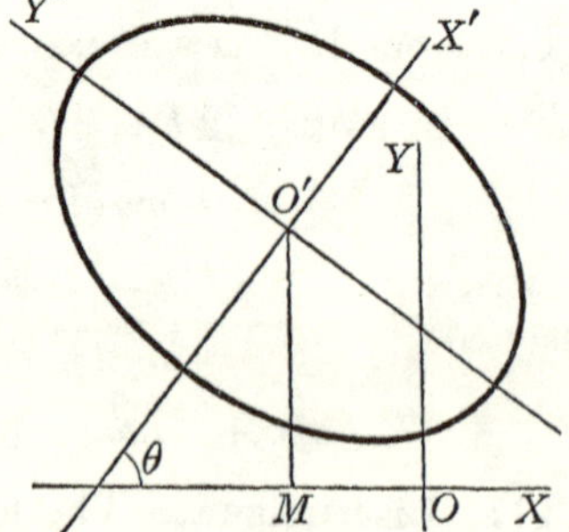

$$a'x^2+b'y^2+c'=0,$$

where $c'=-100$, as shown above, $a'=\frac{1}{2}(75\pm 25)$, and $b'=\frac{1}{2}(75\mp 25)$. Since $a'-b'$ must have the same sign as h, we see that $a'=50$ and $b'=25$, and so the new equation is

$$\frac{x^2}{2}+\frac{y^2}{4}=1,$$

which at once suggests that we lay off the semiaxes on $O'X'$ and $O'Y'$. We then draw the ellipse in the customary way.

2. Given the conic $x^2 - 4xy + 4y^2 - 6x + 2y = 0$, determine the nature of the conic and draw the figure.

Since $ab - h^2 = 0$ the conic is a parabola, and we write the equation in the form

$$(2y - x)^2 - 6x + 2y = 0. \qquad (1)$$

Then $s = 2$, $r = -1$, and we rotate the axes (§ 209) through an angle θ such that $\tan\theta = -r/s = \frac{1}{2}$. Therefore $\sin\theta = 1/\sqrt{5}$, $\cos\theta = 2/\sqrt{5}$, and we write $(2x - y)/\sqrt{5}$ for x and $(x + 2y)/\sqrt{5}$ for y in equation (1). Then we have

$$\left(\frac{5y}{\sqrt{5}}\right)^2 - \frac{10}{\sqrt{5}}x + \frac{10}{\sqrt{5}}y = 0,$$

or

$$\sqrt{5}\,y^2 + 2y = 2x,$$

which may be written in the form

$$\sqrt{5}\,(y - q)^2 = 2(x - p),$$

where $q = -\dfrac{1}{\sqrt{5}}$ and $p = -\dfrac{1}{2\sqrt{5}}$.

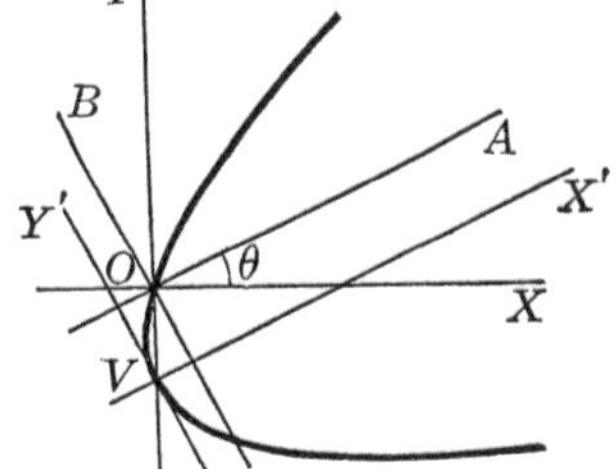

Completing the reduction by moving the origin to the vertex (p, q), that is, to the point $V\left(-\dfrac{1}{2\sqrt{5}}, -\dfrac{1}{\sqrt{5}}\right)$, we have

$$y^2 = \frac{2}{\sqrt{5}}x. \qquad (2)$$

Therefore we draw through V a line VX' making with OX the acute angle $\theta = \tan^{-1}\frac{1}{2}$, and this is the new x axis. The line VY', perpendicular to VX', is the new y axis, and the curve is drawn by plotting a few points, using equation (2) and the new axes.

3. Given the parabola $(4x - 3y)^2 = 250x - 100$, find the coordinates of the vertex.

Take $\tan\theta = \frac{4}{3}$; then $\sin\theta = \frac{4}{5}$ and $\cos\theta = \frac{3}{5}$. Rotating the axes through the angle θ, the new equation is $y^2 + 8y - 6x + 4 = 0$, or $(y + 4)^2 = 6(x + 2)$. Referred to the new axes (§ 122) the vertex is $V(-2, -4)$. If V, referred to the old axes (§ 109), is (a, b), then $a = -2 \cdot \frac{3}{5} - (-4) \cdot \frac{4}{5} = 2$ and $b = -2 \cdot \frac{4}{5} + (-4)\frac{3}{5} = -4$, and hence the vertex of the parabola is the point $(2, -4)$.

213. Illustrative Examples. Degenerate Cases. By computing the value of the discriminant Δ (§ 211) we can always determine in advance whether or not a given conic is degenerate. But when $ab - h^2 \neq 0$, it is usually simpler to begin the work of reduction for a given equation in the manner explained in Ex. 1, § 212. In that case if the equation happens to represent a degenerate conic, the fact will soon become apparent.

If the equation represents two lines, it follows from § 210 that $\Delta = 0$. It is evident that, conversely, if $\Delta = 0$, the steps may be retraced, and the equation represents two lines.

1. Given the equation $y^2 - xy - 6x^2 - 3x + y = 0$, determine the nature of the conic and draw the figure.

Since $ab - h^2 = -6 \times 1 - \frac{1}{4} < 0$, the conic is a hyperbola.

By § 207 the center is $O'(-\frac{1}{5}, -\frac{3}{5})$, and when the origin is moved to O', the new equation is $y^2 - xy - 6x^2 = 0$.

But a polynomial containing terms in x^2, y^2, and xy, and no other terms, can always be factored. The factors may have real or imaginary coefficients.

The equation $y^2 - xy - 6x^2 = 0$ may be written

$$(y - 3x)(y + 2x) = 0,$$

which represents two lines, $y - 3x = 0$ and $y + 2x = 0$. Since their equations have no constant terms, both the lines pass through the new origin O', and the figure is easily drawn.

2. Given the equation $x^2 - 2xy + y^2 - 4x + 4y - 12 = 0$, determine the nature of the conic and draw the figure.

Here $ab - h^2 = 0$ and $\Delta = 0$, and hence the equation is degenerate. The factors of the polynomial are found simply, thus:

$$(x - y)^2 - 4(x - y) - 12 = 0;$$

hence

$$(x - y - 6)(x - y + 2) = 0.$$

This equation represents the parallel lines $x - y - 6 = 0$ and $x - y + 2 = 0$, which are easily drawn.

Exercise 54. Equations of the Second Degree

In each of the following examples determine the nature of the conic and draw the figure:

1. $5x^2 + 2xy + 5y^2 - 12x - 12y = 0.$

2. $x^2 - 5xy + y^2 + 8x - 20y + 15 = 0.$

3. $x^2 + 2xy - y^2 + 8x + 4y - 8 = 0.$

4. $x^2 - 2xy + y^2 + 2x - y - 1 = 0.$

5. $5x^2 + 4xy + 8y^2 - 16x + 8y - 16 = 0.$

6. $9x^2 - 24xy + 16y^2 - 20x + 110y - 75 = 0.$

7. $7x^2 - 17xy + 6y^2 + 23x - 2y - 20 = 0.$

8. $36x^2 + 24xy + 29y^2 - 72x + 126y + 81 = 0.$

9. $25x^2 - 120xy + 144y^2 - 2x - 29y - 1 = 0.$

10. $2x^2 - 6xy + 4y^2 - x + 4y - 3 = 0.$

11. $4x^2 + 4xy + y^2 - 4x - 2y + 1 = 0.$

12. $4x^2 - 12xy + 9y^2 + 6x - 9y - 10 = 0.$

13. $x^2 + xy + y^2 = 1.$

14. Show by means of § 202, (5) and § 208, (1) that if $a + b = 0$, the equation $ax^2 + 2hxy + by^2 + 2gx + 2fy + c = 0$ represents a rectangular hyperbola.

15. Prove the converse of the theorem of Ex. 14.

16. Solve $ax^2 + 2hxy + by^2 + 2gx + 2fy + c = 0$ for x in terms of y, and show that if $abc + 2fgh - af^2 - bg^2 - ch^2 = 0$, the equation represents a pair of lines.

This method of solving the problem of § 210 fails when a has a certain special value. What is that value?

17. Find the foci of the conic $xy - 2y = 5$.

18. Given the conic $x^2 - 2xy + y^2 = 5x$, find the equation of the directrix.

19. Show how to determine whether or not two given equations represent two congruent conics.

AG

214. Conic through Five Points. The general equation

$$ax^2 + 2hxy + by^2 + 2gx + 2fy + c = 0 \qquad (1)$$

has six terms. If we divide by c, the equation becomes

$$px^2 + qxy + ry^2 + sx + ty + 1 = 0. \qquad (2)$$

The equation therefore involves five essential constants, and hence a conic is determined by five conditions.

For example, let us find the equation of the conic passing through the five points $A(1, 0)$, $B(3, 1)$, $C(0, 3)$, $D(-4, -1)$, $E(-2, -3)$.

Consider the two following methods:

1. Let equation (2) represent the conic. Since $A(1, 0)$ is on the conic, $p + s + 1 = 0$; and since B, C, D, E are on the conic, we get four other equations in p, q, r, s, t. From these we find the values of p, q, r, s, t; and substituting these values in (2), we have the desired equation $24x^2 - 73xy + 29y^2 + 33x - 68y - 57 = 0$.

If this method fails, it means that $c = 0$ and that dividing (1) by c was improper. In that case we may divide by some other coefficient.

2. The equation of the line AB is $2y - x + 1 = 0$; that of the line CD is $y - x - 3 = 0$; and that of the pair of lines AB, CD is

$$(2y - x + 1)(y - x - 3) = 0. \qquad (3)$$

Similarly, the equation of the pair of lines AD, BC is

$$(5y - x + 1)(3y + 2x - 9) = 0. \qquad (4)$$

Now form the quadratic equation

$$(2y - x + 1)(y - x - 3) + k(5y - x + 1)(3y + 2x - 9) = 0, \qquad (5)$$

which represents a conic (§ 200). Since the values of x and y which satisfy both (3) and (4) also satisfy (5), the conic (5) passes through the common points of the conics (3) and (4); that is, through the points A, B, C, and D.

We now determine k so that (5) also passes through $E(-2, -3)$. Substituting -2 for x and -3 for y in (5) we find $k = -\frac{1}{22}$. Putting $-\frac{1}{22}$ for k in (5) and simplifying, we have the desired equation $24x^2 - 73xy + 29y^2 + 33x - 68y - 57 = 0$.

PROBLEM. POLES AND POLARS

215. *To find the equation of the chord of contact of the tangents from a point $P_1(x_1, y_1)$ to the parabola $y^2 = 4px$.*

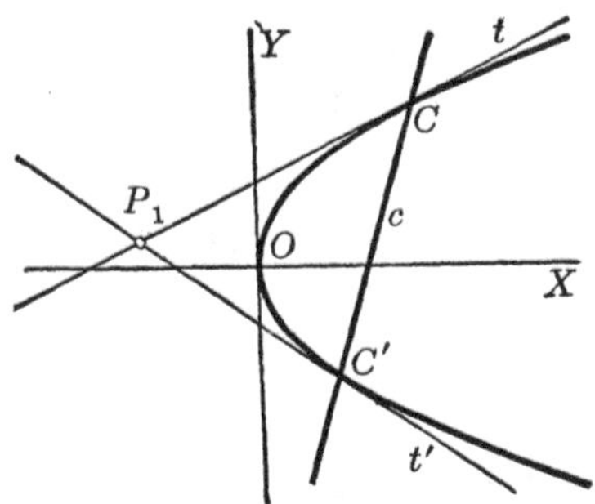

Solution. Let $C(h, k)$ and $C'(h', k')$ be the points of contact of the tangents t and t' from P_1 to the parabola.

Then the equations (§ 127) of the tangents t and t' are

$$ky = 2p(x + h) \quad \text{and} \quad k'y = 2p(x + h').$$

And since t and t' pass through $P_1(x_1, y_1)$, we have

$$ky_1 = 2p(x_1 + h) \tag{1}$$

and

$$k'y_1 = 2p(x_1 + h'). \tag{2}$$

But (1) and (2) show that the points (h, k), (h', k') are on the line

$$\mathbf{y_1 y = 2p(x + x_1),}$$

and therefore this is the required equation.

This equation is of the same form as that of the tangent at any point on the curve. In fact, as $P_1(x_1, y_1)$ approaches the curve at P', the chord of contact approaches the tangent at P'.

216. Pole and Polar. The point P_1 is called the *pole* of the line c with respect to the conic, and the line c is called the *polar* of the point P_1.

217. Poles and Polars for Other Conics. The method of § 215 is obviously applicable to the other conics, the equation of the polar in each case having the same form as the equation of the tangent at a point (x_1, y_1) on the curve. Hence the equations of the polars of the point $P_1(x_1, y_1)$ with respect to the other conics are as follows:

The circle $x^2+y^2=r^2$, $\qquad x_1x+y_1y=r^2.$

The ellipse $\frac{x^2}{a^2}+\frac{y^2}{b^2}=1$, $\qquad \frac{x_1x}{a^2}+\frac{y_1y}{b^2}=1.$

The hyperbola $\frac{x^2}{a^2}-\frac{y^2}{b^2}=1$, $\qquad \frac{x_1x}{a}-\frac{y_1y}{b^2}=1.$

The polar of a real point with respect to a conic whose equation has real coefficients is a real line, although the tangents to a conic from a point inside meet the conic in imaginary points.

THEOREM. RECIPROCAL RELATION OF POLARS

218. *If the polar of the point $P_1(x_1, y_1)$ with respect to any conic passes through the point $P_2(x_2, y_2)$, the polar of the point P_2 with respect to the conic passes through the point P_1.*

Proof. We shall prove this theorem only in the case of the circle, but the method of proof evidently applies to all other conics.

The polar of the point $P_1(x_1, y_1)$ with respect to the circle $x^2+y^2=r^2$ is

$$x_1x+y_1y=r^2. \qquad \text{§ 217}$$

By hypothesis this line passes through the point $P_2(x_2, y_2)$; that is

$$x_1x_2+y_1y_2=r^2.$$

But this equation shows that $P_1(x_1, y_1)$ is on the line $x_2x+y_2y=r^2$, which is the polar of P_2.

It will be instructive for the student to draw for each kind of conic a figure showing the relations described in this theorem.

Exercise 55. Poles and Polars

Find the equation of the polar of P with respect to each of the following conics, and in each case draw the figure, showing both tangents from P, if real, and the polar:

1. $4x^2 + 9y^2 = 36$, $P(-2, -2)$.

2. $16x^2 - 9y^2 = 144$, $P(6, 3)$.

3. $4x^2 + y^2 = 254$, $P(2, 3)$.

4. $x^2 + y^2 = 169$, $P(-5, 12)$.

5. If O is the center of the circle $x^2 + y^2 = r^2$, and the polar of $P_1(x_1, y_1)$ with respect to the circle cuts OP_1 at Q, then the polar is perpendicular to OP_1, and $OP_1 \cdot OQ = r^2$.

6. Give a geometric construction for the polar of a given point with respect to a given circle, the point being either inside or outside the circle.

7. The points P_1 and Q in Ex. 5 divide the diameter of the circle harmonically.

That is, the points divide the diameter of the circle internally and externally in the same ratio.

8. Show that the polar of the focus of a given parabola, ellipse, or hyperbola is in each case the directrix, and investigate the existence of the polar of the center of a circle.

9. Find the coordinates of the pole of the line $4x - 2y = 3$ with respect to the ellipse $4x^2 + y^2 = 9$.

If the pole is (x_1, y_1), the polar is $4x_1x + y_1y = 9$. By comparing these coefficients with those of $4x - 2y = 3$ it is easy to find x_1 and y_1.

10. The diameter which bisects a chord of a parabola passes through the pole of the chord.

11. Find the points of contact of the tangents from the point (1, 2) to the circle $x^2 + y^2 = 1$.

12. The polars with respect to a given conic of all points on a straight line pass through the pole of that line.

Exercise 56. Review

1. Find the coordinates of the vertex of the parabola $x^2 - 2xy + y^2 - 6x - 10 = 0$.

2. Show that the equation $x^2 - 6xy + 9y^2 = 25$ represents a pair of parallel straight lines, and draw these lines.

3. Draw the conic $x^2 - 5xy + 6y^2 = 0$.

4. If the equation $xy + ax + by + c = 0$ represents two lines, one of these lines is parallel to OX and the other to OY.

5. Two conics intersect in four points, real or imaginary.

6. Find the equation of the conic passing through the points $(1, -1)$, $(2, 0)$, $(1, 1)$, $(0, 0)$, $(0, -1)$.

7. Find the equation of the conic passing through the point $A(2, -2)$ and through the four common points of the conics $x^2 + xy - 2y^2 + 6x - 1 = 0$ and $2x^2 - y^2 - x - y = 0$.

Show that the conic $x^2 + xy - 2y^2 + 6x - 1 + k(2x^2 - y^2 - x - y) = 0$ passes through the common points of the two given conics, and then find k so that this conic passes through A.

8. Find the equation of the real axis of the hyperbola $5x^2 - 24xy - 2y^2 - 4x + 3y = 7$.

The axis passes through the center and makes the angle θ with OX.

9. Find the equation of the tangent drawn to the parabola $x^2 - 2xy + y^2 - 4x + y - 10 = 0$ at the vertex.

10. The equation $x^{\frac{1}{2}} + y^{\frac{1}{2}} = a^{\frac{1}{2}}$ represents a parabola which is tangent to both axes.

11. The discriminant $abc + 2fgh - af^2 - bg^2 - ch^2$ can be written in the form of a determinant, as follows:

$$\begin{vmatrix} a & h & g \\ h & b & f \\ g & f & c \end{vmatrix}$$

Ex. 11 should be omitted by those who have not studied determinants.

12. Given the conic $ax^2 + 2hxy + by^2 + 2gx + 2fy + c = 0$, find the slope at any point (x', y') on the conic.

CHAPTER XI

POLAR COORDINATES

219. Polar Coordinates. We have hitherto located points in the plane by means of rectilinear coordinates. We shall now explain another important method of locating points.

Let O be a fixed point and OX a fixed line. Then any point P in the plane is determined by its distance OP from O and by the angle θ from OX to OP. The distance OP is denoted by ρ.

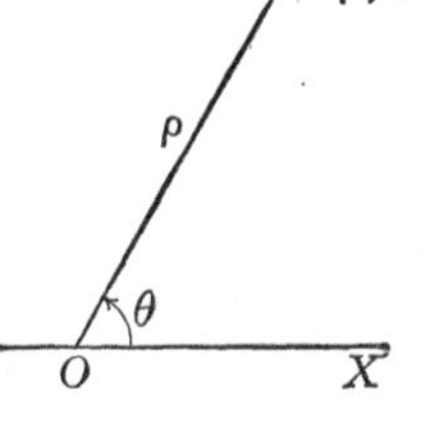

The magnitudes ρ and θ are called the *polar coordinates* of P, ρ being called the *radius vector*, and θ the *vectorial angle*. The point O is called the *pole* or *origin*, and the line OX is called the *polar axis*. Polar coordinates are denoted thus: (ρ, θ).

220. Convention of Signs. We regard θ as positive when it is measured counterclockwise and as negative when measured clockwise. We regard ρ as positive when it is measured from O along the terminal side of θ and as negative when measured in the opposite direction.

Thus, in this figure we may locate P if it is given that $\theta = 30°$ and $\rho = 7$. The same point P is located by $\theta = 210°$ and $\rho = -7$, as well as in any one of several other ways. But given any point P, we shall, whenever we are free to choose, always take θ as the positive angle XOP, the corresponding value of ρ being therefore positive.

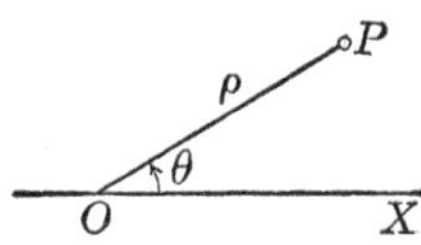

221. Special Equations. We shall now consider two important equations and their graphs.

1. Consider the graph of the equation $\rho = 2\,a \cos\theta$.

This equation represents a circle with diameter $2\,a$, the circle passing through the pole and having its center on the polar axis.

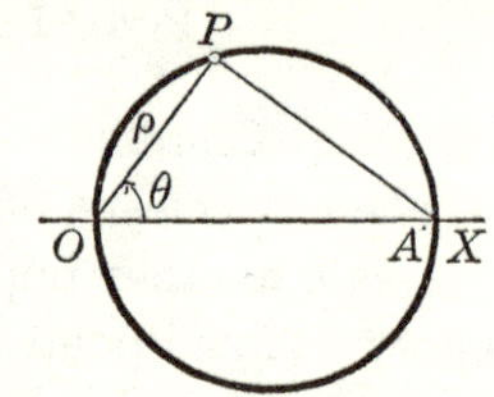

For if $P(\rho, \theta)$ is any point on the circle, $OP = \rho$ and the angle $OPA = 90°$. Hence $\cos\theta = \rho/2\,a$, or $\rho = 2\,a\cos\theta$, and this equation is obviously not true unless P is on the circle. This equation is called the *polar equation* of this circle.

2. Consider the graph of $\rho = 2\,a \sin\theta$.

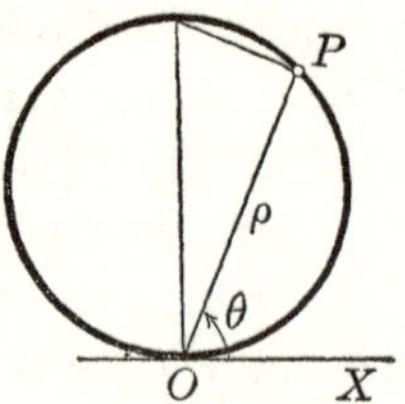

This equation represents a circle with diameter $2\,a$, the circle being tangent to the polar axis at O.

This is apparent when we consider that in this case we have $\cos(90° - \theta) = \rho/2\,a$; that is, that $\rho = 2\,a \sin\theta$.

222. Relations of Rectangular and Polar Coordinates. If P is any point in the plane and has the rectangular coordinates (x, y) and the polar coordinates (ρ, θ), then

$$x = \rho\cos\theta$$

and

$$y = \rho\sin\theta.$$

Hence, if the equation of a curve is known in rectangular coordinates, the polar equation of the curve may be found by making the above substitution.

To change an equation from polar to rectangular coordinates we must substitute the following values:

$$\rho = \sqrt{x^2 + y^2},\quad \cos\theta = x/\sqrt{x^2 + y^2},\quad \sin\theta = y/\sqrt{x^2 + y^2}.$$

Exercise 57. Polar Coordinates

Plot the following points:

1. $(4, 45°)$, $(2, 90°)$, $(3, 60°)$, $(-3, 60°)$, $(3, 240°)$, $(1, 0°)$.

2. $(5, 120°)$, $(2, 180°)$, $(-2, 180°)$, $(-6, 330°)$, $(-3, -30°)$.

3. $(8, 390°)$, $(-8, 210°)$, $(7, 15°)$, $(-7, -240°)$, $(5, 120°)$.

4. Plot the following points, and give for each a pair of coordinates in which the radius vector is positive: $(-4, 225°)$, $(-10, 300°)$, $(-4, 75°)$, $(-9, -60°)$, $(-8, 36°)$.

5. What is the locus of all points for which $\rho = 10$? What is the equation in polar coordinates of a circle having a radius r and its center at the origin?

6. What is the locus of all points for which $\theta = 70°$? What is the equation in polar coordinates of a line passing through the origin?

7. Draw the line l perpendicular to the polar axis at Q so that $OQ = a$. Let $P(\rho, \theta)$ be any point on l, draw ρ and θ, and show that the equation of l is $\rho \cos \theta = a$.

8. The equation of the line parallel to the polar axis and distant a from it is $\rho \sin \theta = a$.

Change the following to equations in polar coordinates:

9. $y^2 = 12x$.

10. $x^2 + y^2 = 4x$.

11. $x^2 - y^2 = 20$.

12. $4x^2 + y^2 = 4$.

13. $x^2 - 2xy + y^2 = x - 4$.

14. $xy = 7$.

15. $2x^2 - xy + y^2 - 6x + y = 1$.

16. $(x^2 + y^2)^2 = 4(x^2 - y^2)$.

Change the following to equations in rectangular coordinates:

17. $\rho = 2a \cos \theta$.

18. $\rho \sin \theta = 10$.

19. $\rho = 4 \sin 2\theta$.

20. $\rho^2 \cos 2\theta = -1$.

21. $\rho \cos (\theta - 30°) = 1$.

22. $\rho (\sin \theta + 2 \cos \theta) = 6$.

In Ex. 19 recall the fact that $\sin 2\theta = 2 \sin \theta \cos \theta$.

223. Graphs in Polar Coordinates. To plot the graph of an equation in polar coordinates we assign values to θ, compute the corresponding values of ρ, and from these values locate a sufficient number of points of the curve.

1. Plot the graph of the equation $\rho = \dfrac{10}{2 - \cos\theta}$.

$\theta =$	0°	30°	60°	90°	120°	150°	180°
$\rho =$	10. 0	8.8	6.7	5.0	4.0	3.5	3.3

Since $\cos(-\theta) = \cos\theta$, it appears that a negative value of θ, say $\theta = -k$, leads to the same value of ρ as the positive value $\theta = k$, and hence the curve is symmetric with respect to OX. It is only necessary to assign to θ values between 0° and 180°, draw the curve above OX, and then draw below OX a curve that is symmetric to it.

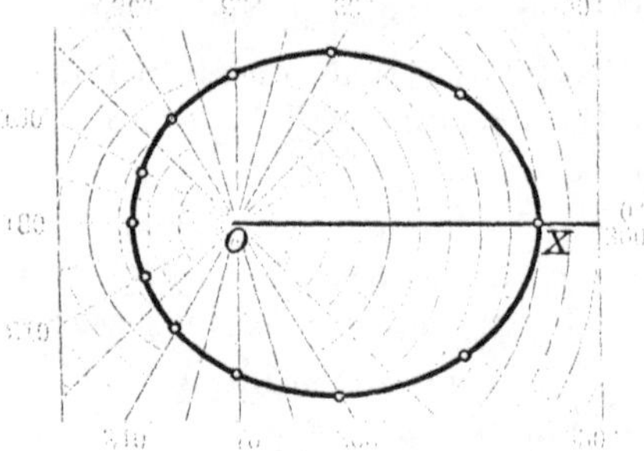

2. Plot the graph of the equation $\rho = 10 \sin 2\theta$.

$\theta =$	0°	15°	30°	45°	60°	75°	90°
$\rho =$	0	5.0	8.7	10.0	8.7	5.0	0

It is left to the student to assign values to θ in the other three quadrants and to complete the graph.

If the student has forgotten the trigonometric functions of the familiar angles 0°, 30°, 45°, ···, he may refer to page 285.

Paper conveniently ruled for plotting the graphs of equations in polar coordinates can be obtained from stationers.

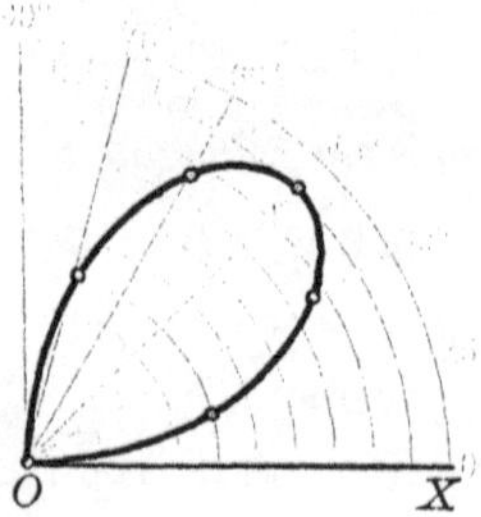

224. Polar Equation of a Conic. Let F be the focus, l the directrix, and e the eccentricity of any conic, and denote by $2p$ the distance from l to F. Draw RP perpendicular to FX.

Let $P(\rho, \theta)$ be any point on the conic. Then the equation of the conic in polar coordinates is found from the definition of a conic (§ 112) as the locus of a point P which moves so that the ratio of its distances from F and l is a constant, this constant being the eccentricity e.

Then $$\frac{FP}{QP} = e,$$

that is, $$FP = e \cdot SR$$

$$= e(SF + FR);$$

whence $$\rho = e(2p + \rho \cos \theta),$$

which is the desired equation. Solving this equation for ρ in terms of θ, we have

$$\rho = \frac{2ep}{1 - e\cos\theta}.$$

This is the equation (§ 113) of the parabola, ellipse, or hyperbola, according as $e = 1$, $e < 1$, or $e > 1$.

In the figure we have taken the focus to the right of the directrix. Hence the equation obtained assumes that the left-hand focus is the pole in the case of the ellipse, but that the right-hand focus is the pole for the hyperbola (§§ 138, 167). In the case of the ellipse or the hyperbola, if the other focus is taken as the pole, the student may show that the resulting equation is

$$\rho = \frac{2ep}{1 + e\cos\theta},$$

which differs from the equation given above only in the sign of one of the quantities involved in the fraction.

Since $e = 1$ in the case of the parabola, it is evident that the equation of this curve is

$$\rho = \frac{2p}{1 - \cos\theta}.$$

Exercise 58. Polar Graphs

Plot the graphs of the following equations:

1. $\rho = 10 \sin \theta$.
2. $\rho = 10 \cos \theta$.
3. $\rho = 1 + \cos \theta$.
4. $\rho = 6(1 - \cos \theta)$.
5. $\rho = 1 + \sin \theta$.
6. $\rho = 4 \cos 2\theta$.
7. $\rho = 6 \sin 3\theta$.
8. $\rho = 6(\cos \theta + \sin \theta)$.
9. $\rho = 4 \cos \theta - 2 \sin \theta$.
10. $\rho^2 = 16 \sin \theta$.
11. $\rho^2 = 8 \cos 2\theta$.
12. $\rho(\cos \theta - 3 \sin \theta) = 8$.

13. Find the equation of the ellipse in polar coordinates, taking the right-hand focus as the pole.

14. Find the equation of the hyperbola in polar coordinates, taking the left-hand focus as the pole.

15. The equation of the ellipse, when the center of the ellipse is the pole and the major axis is the polar axis, is

$$\rho^2 = \frac{a^2b^2}{b^2 \cos^2\theta + a^2 \sin^2\theta}.$$

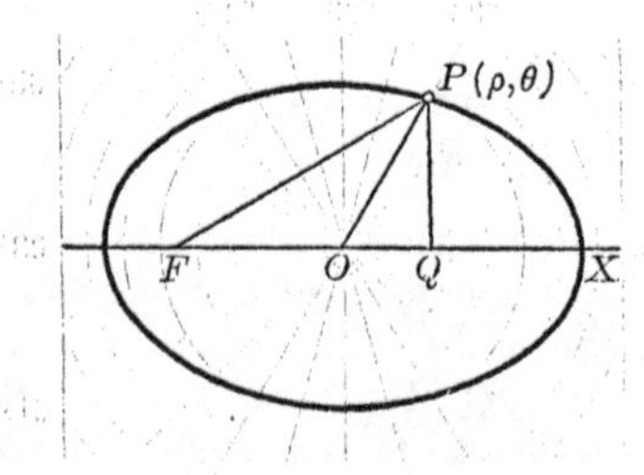

This may be shown most simply by changing the equation in rectangular coordinates to one in polar coordinates (§ 222). For a direct derivation draw the triangle FPO, where O is the center and $P(\rho, \theta)$ is any point on the curve. Draw PQ perpendicular to OX. Then we have $FO = \sqrt{a^2 - b^2}$ (§ 140), $OQ = \rho \cos \theta$, $QP = \rho \sin \theta$, $FP = a + e \cdot OQ = a + e\rho \cos \theta$ (§ 143), and, finally, $\overline{FQ}^2 + \overline{QP}^2 = \overline{FP}^2$.

16. Find the equation of the hyperbola when the center is the pole and the real axis is the polar axis.

17. Given a circle with radius r and the pole O on the circle; if the center is the point (r, α), the equation of the circle is $\rho = 2r \cos(\theta - \alpha)$.

18. Draw the graph of the equation $2\rho \cos \theta = 2 + \rho$.

19. Where are all points for which $\theta = 0°$? for which $\theta = 180°$? for which $\rho = 0$?

20. Draw the graph of the equation $\rho = 10 \sin \frac{1}{2} \theta$.

21. Draw the graph of the equation $\rho = 8/(2 - \cos \theta)$.

22. Draw the graph of the equation $\rho = 8/(1 - 2 \cos \theta)$.

23. From the polar equation of the parabola show that the focal width is $4p$.

24. From their polar equations find the focal widths of the ellipse and the hyperbola.

25. Find the intercepts on the polar axis of the ellipse, hyperbola, and parabola.

26. Change the equation $\rho^2 = 10 \cos 2\theta$ to an equation in rectangular coordinates.

Since $\cos 2\theta = \cos^2\theta - \sin^2\theta$, the equation may be written in the form $\rho^4 = 10(\rho^2 \cos^2\theta - \rho^2 \sin^2\theta)$.

27. Draw the graph of the equation $(x^2 + y^2)^2 = 10x^2 - 10y^2$ by first changing the equation to an equation in polar coordinates.

Which of the two equations appears the easier to plot?

28. How does the graph of the equation $\rho = a \cos \theta$ differ from that of $\rho^2 = a\rho \cos \theta$, obtained by multiplying the members of $\rho = a \cos \theta$ by ρ?

The equation $\rho^2 = a\rho \cos\theta$, or $\rho^2 - a\rho \cos\theta = 0$, is degenerate. Consider the factors of the left-hand member.

29. Draw the graphs of $\rho = a \sin \theta$ and $\rho\theta = a\theta \sin \theta$.

It should be observed that the second equation is degenerate.

30. Determine the nature of the graph of the general equation of the first degree in polar coordinates.

31. Find the equation of the locus of the mid points of the chords of a circle which pass through a point O on the circle.

32. Draw a line from the focus F to any point Q on an ellipse, prolong FQ to P making $FQ = QP$, and find the equation of the locus of P.

33. A straight line OB begins to rotate about a fixed point O at the rate of 10° per second, and at the same instant a point P, starting at O, moves along OB at the rate of 5 in. per second. Find the equation of the locus of P.

34. The sum of the reciprocals of the segments FP and FQ into which the focus F divides any focal chord of a conic is constant.

If the angle $XFQ = \theta$, then $PFX = 180° + \theta$, and the corresponding values of ρ in the polar equation of the conic are FQ and FP, where F is taken as the pole and FX as the polar axis.

35. The product of the segments into which the focus divides a focal chord of a conic varies as the length of the chord; that is, the ratio of the product of the segments to the length of the chord is constant.

36. The sum of the reciprocals of two perpendicular focal chords of a conic is constant.

37. If PFP' and QFQ' are perpendicular focal chords of a conic, then $\dfrac{1}{PF \cdot FP'} + \dfrac{1}{QF \cdot FQ'}$ is constant.

38. Find the locus of the mid point of a variable focal radius of a conic.

39. Find the equation of a given line in polar coordinates.

Denote by p the perpendicular distance from the pole to the line, and by β the angle from OX to p. Take any point $P(\rho, \theta)$ on the line, and show that $\rho \cos(\theta - \beta) = p$.

40. The equation of the circle with center $C(a, \alpha)$ and radius r is $\rho^2 + a^2 - 2a\rho\cos(\theta - \alpha) = r^2$.

Draw the coordinates of C and of any point $P(\rho, \theta)$ on the circle, and draw the radius CP.

41. Find the polar equation of the locus of a point which moves so that its distance from a fixed point exceeds by a constant its distance from a fixed line. Show that the locus is a parabola having the fixed point as focus, and find the polar equation of the directrix.

CHAPTER XII

HIGHER PLANE CURVES

225. Algebraic Curve. If the equation of a curve referred to the rectilinear coordinates x and y involves only algebraic functions of x and y (§ 58), the curve is called an *algebraic curve.*

We have studied the algebraic curves represented by equations of the first and second degrees. It is a much more difficult matter to discover all the kinds of curves represented by equations of the third degree and by equations of degree higher than the third.

Newton was the first to make a searching investigation of the curves represented by equations of the third degree, and according to his classification there are seventy-eight kinds.

In the first part of this chapter we shall describe a few curves of special interest the equations of which are of the third and fourth degrees.

226. Transcendent Curve. If the equation of a curve referred to rectilinear coordinates involves transcendent functions of x or y (§ 59), the curve is called a *transcendent curve.* Such, for example, are the graphs of the equations involving $\sin x$, $\log x$, and 10^x. We shall examine a few of these curves in this chapter.

As an example of a useful formula containing a constant raised to a variable power, it is easily seen that y, the amount of $1 at 6% interest compounded annually for x years, is found from the equation $y = 1.06^x$, and that the graph of this equation is a transcendent curve.

227. Cissoid of Diocles. Let OK be a diameter of a circle with radius r, and let KH be tangent to this circle at the point K. Let the line OR from O to KH cut the circle at S, and take OP on OR so that $OP = SR$. The locus of the point P as OR turns about O is called a *cissoid*.

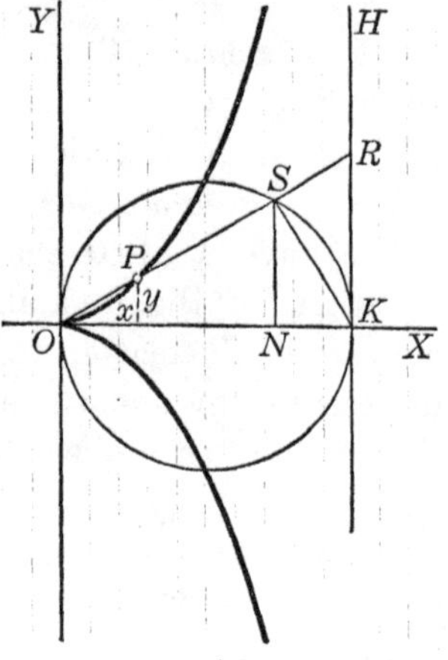

To find the equation of the cissoid referred to the rectangular axes OX and OY, draw the coordinates x, y of P, draw SK, and draw SN perpendicular to OX.

Then $$\frac{y}{x} = \frac{NS}{ON}. \qquad (1)$$

We have now only to find each of the lengths NS and ON in terms of x and y, which may be done as follows:

Since $SR = OP$,

we have $NK = x$

and $ON = 2r - x$.

In the right triangle OSK it is evident from elementary geometry that NS is the mean proportional between ON and NK, and hence we have

$$\begin{aligned} NS &= \sqrt{ON \cdot NK} \\ &= \sqrt{(2r - x)x}. \end{aligned}$$

Substituting these values of ON and NS in (1), and squaring to remove radicals, we have

$$y^2 = \frac{x^3}{2r - x}.$$

This is one of the simplest curves having equations of the third degree in x and y.

228. Shape of the Cissoid. From the equation of the cissoid we see that the curve is symmetric with respect to OX (§ 43), has real points only when $0 \leqq x < 2r$, and approaches the vertical asymptote $x = 2r$ (§ 47).

Also, as P approaches O the secant OPR approaches a horizontal position of tangency at O (§ 124).

The student should note that $y = \pm r$ when $x = r$. That is, the cissoid bisects the semicircumferences above and below OX.

This curve was thought by the ancients to resemble an ivy vine, and hence its name, the word *cissoid* meaning "ivylike."

229. Duplication of the Cube. The cissoid was invented by Diocles, a Greek mathematician of the second century B.C., for a solution of the problem of finding the edge of the cube having twice the volume of a given cube.

To duplicate the cube, draw a line through C, the center of a circle of radius r, perpendicular to OX, and on it take $CB = 2r$. Draw BK cutting the cissoid in Q. Then, since $CK = \frac{1}{2}CB$, $EK = \frac{1}{2}EQ$. But from the equation of the cissoid we have

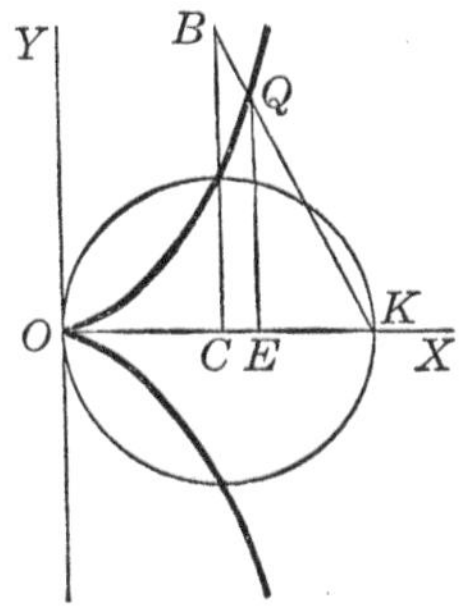

$$\overline{EQ}^2 = \frac{\overline{OE}^3}{EK} = \frac{\overline{OE}^3}{\frac{1}{2}EQ};$$

whence $\overline{EQ}^3 = 2\,\overline{OE}^3.$

Hence if a is the edge of a given cube, construct b so that $OE : EQ = a : b$. Then $\overline{OE}^3 : \overline{EQ}^3 = a^3 : b^3$, and since $\overline{EQ}^3 = 2\,\overline{OE}^3$, we have $b^3 = 2a^3$, and the problem is solved.

In like manner, by taking $CB = nr$, we can find the edge of a cube n times the given cube in volume.

The student will readily imagine, however, that the instruments of geometric construction with which he is familiar — the straight edge and compasses — will not suffice for drawing the cissoid.

230. Conchoid of Nicomedes. If a line OP revolves about a fixed point O and cuts a fixed line l in the point R, the locus of a point P such that RP has a constant length c is called a *conchoid*.

To find the equation of the conchoid, take rectangular axes through the fixed point O, so that OY is perpendicular to the line l. Denote by a the distance from O to l, and take $P(x, y)$ any point on the curve. Draw PB perpendicular to OX and cutting l in the point M.

Then, whether the distance RP, or c, is laid off above l or below l, we have the proportion

$$OP : RP = BP : MP;$$

that is, $$\sqrt{x^2 + y^2} : c = y : y - a,$$

or $$(x^2 + y^2)(y - a)^2 = c^2y^2.$$

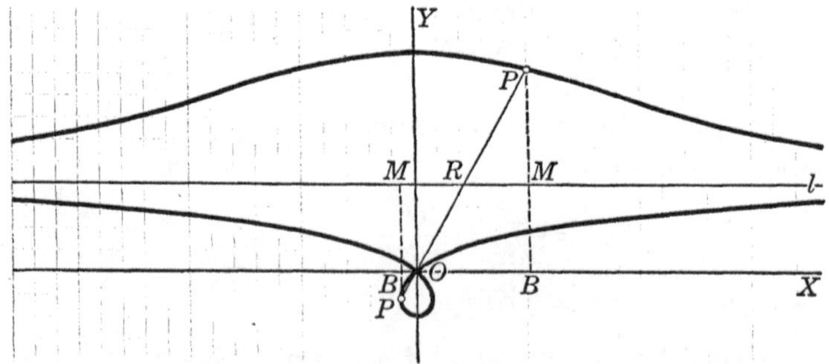

When P is below l, the student may find it difficult to see that $MP = y - a$. But this is simple, for $BM = a$, and hence $MB = -a$.

The above figure shows the curve for the case $c > a$. The student should draw the curve for the cases $c < a$ and $c = a$.

The equation of this curve is of the fourth degree, as shown above.

This curve was thought by the ancients to represent the outline of a mussel shell, and hence its name, the word *conchoid* meaning "shell-like."

231. Shape of the Conchoid. The conchoid is symmetric with respect to OY. Furthermore, if we solve for x^2 the equation of the conchoid, we have

$$x^2 = \frac{c^2y^2 - y^2(y-a)^2}{(y-a)^2},$$

whence we see that $y = a$, which is the line l, is a horizontal asymptote for both branches of the curve, the one above l and the other below l.

232. Trisection of an Angle. The conchoid was invented by Nicomedes, a Greek mathematician of the second century B.C., for the purpose of solving the problem of the duplication of the cube. It is also applicable to the trisection of an angle, a problem not less celebrated among the ancients.

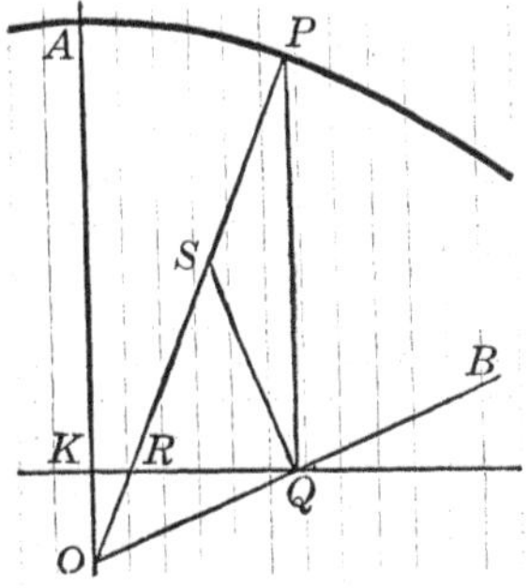

To trisect any angle, as AOB, lay off any length OQ on OB. Draw QK perpendicular to OA, and lay off $KA = 2\,OQ$. Now draw a conchoid with O as the fixed point mentioned in the definition, QK as the fixed line, and KA as the constant length c.

At Q erect a perpendicular to QK cutting the curve in the point P. Then OP trisects the angle AOB.

To prove this, bisect PR at S.

Then $$QS = SP = \tfrac{1}{2}RP = \tfrac{1}{2}KA = OQ,$$

and $$\angle SOQ = \angle QSO = 2\angle QPS = 2\angle AOP.$$

Therefore $$\angle AOP = \tfrac{1}{3}\angle AOB.$$

Since AOB is any angle whatever, and $\angle AOP$ is one third of it, the problem is solved.

233. Lemniscate of Bernoulli. The locus of the intersection of a tangent to a rectangular hyperbola with the perpendicular from the center to the tangent is called the *lemniscate of Bernoulli.*

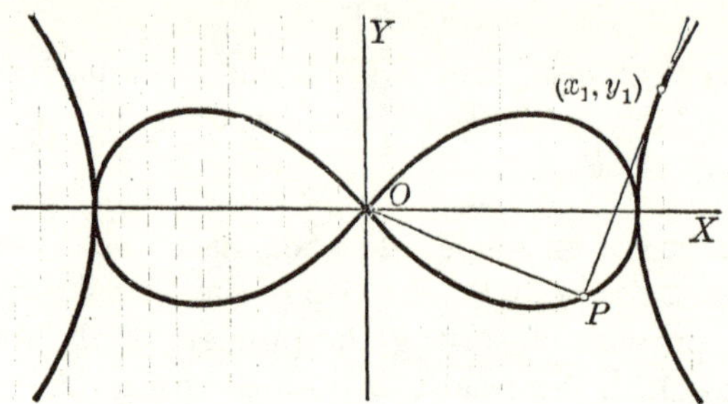

To find the equation of the lemniscate of Bernoulli, we have first the rectangular hyperbola $x^2 - y^2 = a^2$. We have also the equation of the tangent at (x_1, y_1),

$$x_1x - y_1y = a^2, \tag{1}$$

and the equation of the perpendicular to the tangent from O,

$$x_1y + y_1x = 0. \tag{2}$$

If we regard (1) and (2) as simultaneous, x and y are the coordinates of the intersection of (1) and (2).

Finding x_1 and y_1 in terms of x and y, we have

$$x_1 = a^2x/(x^2 + y^2), \qquad y_1 = -a^2y/(x^2 + y^2).$$

But (x_1, y_1) is on the hyperbola, and hence $x_1^2 - y_1^2 = a^2$. Substituting, we have the desired equation

$$\mathbf{(x^2 + y^2)^2 = a^2(x^2 - y^2).}$$

This curve was invented by Jacques Bernoulli (1654–1705). Fagnano discovered (1750) its principal properties, but the analytic theory is due chiefly to Euler. The name means "ribboned."

234. Cycloid. The locus of a point on a circle as the circle rolls along a straight line is called a *cycloid.*

To find the equation of the cycloid, take OX as the x axis, and take the point O, where the curve meets OX, as the origin.

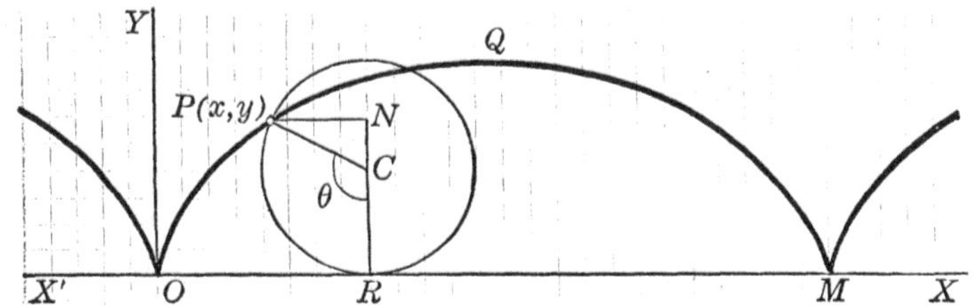

Let $P(x, y)$ be any point on the curve, r the radius of the rolling circle, and θ the angle PCR, measured in radians. Then arc $PR = r\theta$, and arc PR is equal to OR, the line over which the circle has rolled since P was at O.

Then $$x = OR - PN = r\theta - r\sin\theta,$$

and $$y = RC - NC = r - r\cos\theta.$$

That is, $$\boldsymbol{x = r(\theta - \sin\theta)}$$

and $$\boldsymbol{y = r(1 - \cos\theta).}$$

Students not familiar with radian measure should consult page 285.

It is not difficult to eliminate the variable θ from these equations and thus obtain the equation of the curve in the variables x and y. But the resulting equation is not nearly so simple as these two equations, and hence the latter are commonly used in studying the curve.

Eliminating θ from the above equations we have

$$x = r\cos^{-1}\frac{r-y}{y} \mp \sqrt{2ry - y^2}.$$

The curve consists of an unlimited number of arches, but a single arch is usually termed a cycloid. The name means "circle-like."

235. Properties of the Cycloid. To find the x intercept OM, let $y=0$ in the equation $y=r(1-\cos\theta)$ obtained in § 234. Then $\theta=0,\ 2\pi,\ 4\pi,\cdots$, and $x=0,\ 2\pi r,\ 4\pi r,\cdots$. Hence the cycloid meets the x axis in the points $(0, 0), (2\pi r, 0), (4\pi r, 0), \cdots$, and the length of OM is $2\pi r$. The value of y is obviously greatest when $\cos\theta$ is least, that is, when $\theta=\pi$, in which case $y=2r$.

These facts are also easily seen geometrically.

The invention of the cycloid is usually ascribed to Galileo. Aside from the conics, no curve has exercised the ingenuity of mathematicians more than the cycloid, and their labors have been rewarded by the discovery of a multitude of interesting properties. Thus the length of the arch OQM (§ 234) is eight times the radius, and the area OQM is three times the area of the generating circle.

Also, the cycloid is the curve of quickest descent, the *brachistochrone curve*; that is, a ball rolls from a point P to a lower point Q, not vertically below P, in less time on an inverted cycloid than on any other path.

236. Parametric Equations. When the coordinates x and y of any point P on a curve are given in terms of a third variable θ, as in § 234, θ is called a *parameter*, and the equations which give x and y in terms of θ are called the *parametric equations* of the curve.

To plot a curve when its parametric equations are given, assign values to the parameter, compute the corresponding values of x and y, and thus locate points.

For an easy example of parametric equations of a curve, draw a circle with radius r and center at $(0, 0)$. Draw the coordinates x, y of any point P on the circle. Denoting the angle XOP by θ, we have $x=r\cos\theta$ and $y=r\sin\theta$ as parametric equations of the circle.

In the study of the ellipse (§ 163) we expressed x and y in terms of the eccentric angle ϕ for the point $P(x, y)$ as $x=a\cos\phi$ and $y=b\sin\phi$; hence these are parametric equations of the ellipse.

The student may prove that the parametric equations $x=(t-1)^2$ and $y=2(t-1)$ represent the parabola $y^2=4x$.

237. Hypocycloid. The locus of a point on a circle which rolls on the inside of a fixed circle is called a *hypocycloid.*

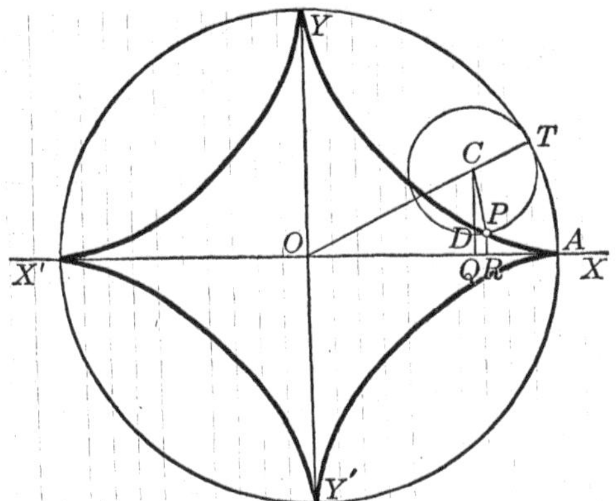

Taking the origin at the center O of the fixed circle and the x axis through a point A where the locus meets the fixed circle, let $P(x, y)$ be any point on the locus.

Draw the line of centers OC, and let the ratio $OT : CT = n$. Then if $CT = r$, we have $OT = nr$ and $OC = (n-1)r$.

Let $\angle AOT = \theta$ radians. Then the arc PT is equal to the arc AT over which it has rolled, that is, $r \times \angle PCT = nr \cdot \theta$.

Hence $\qquad \angle PCT = n\theta.$

Also, $\qquad \angle QCP = \pi - \angle OCD - \angle PCT$

$$= \tfrac{1}{2}\pi - (n-1)\theta.$$

Then
$$x = OQ + QR$$
$$= OC\cos\theta + CP\sin QCP,$$
$$y = QC - DC$$
$$= OC\sin\theta - CP\cos QCP.$$

Therefore
$$\boldsymbol{x = (n-1)r\cos\theta + r\cos(n-1)\theta,}$$
$$\boldsymbol{y = (n-1)r\sin\theta - r\sin(n-1)\theta.}$$

If $n = 4$, the hypocycloid has four *cusps*, as in the figure.

Exercise 59. Higher Plane Curves

1. Find the polar equation of the cissoid.

This may be done by the substitution of $\rho \cos \theta$ for x and of $\rho \sin \theta$ for y, or it may be done independently from the figure of § 227 by observing that $\rho = OP = OR - OS$ and finding OR and OS in terms of ρ and θ.

2. Find the polar equation of the conchoid.

3. Find the polar equation of the lemniscate of Bernoulli.

4. Draw a circle with radius r, take the origin O on the circle, and take the polar axis through the center. A line rotating about O cuts the circle at the point R, and a point P is taken on OR so that the length RP is a constant a. Find the equation of the locus of P. Sketch the curve for the three cases where $a < 2r$, $a = 2r$, and $a > 2r$.

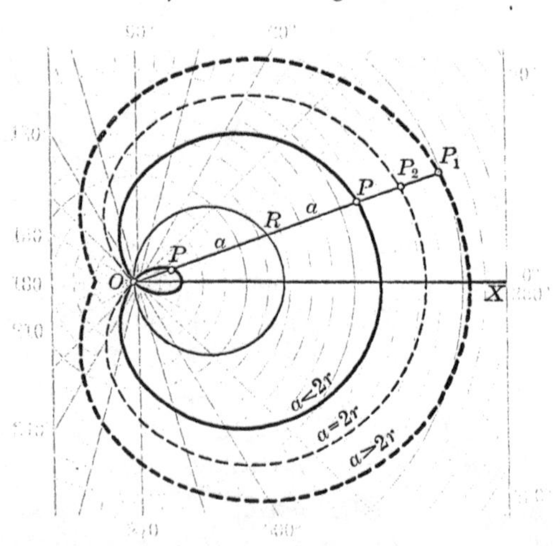

The polar equation is easily found if it is noted that $OR = 2r \cos \theta$.

This locus is called a *limaçon* (from the Latin *limax*, "a snail"). When $a = 2r$ it is called a *cardioid* (meaning "heart-shaped").

5. If F and F' are two fixed points, find the equation of the locus of a point P when $F'P \cdot FP = a^2$ where a is a constant.

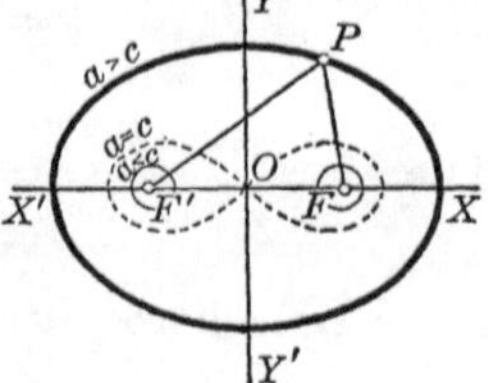

This locus is called *Cassini's oval*, named from Giovanni Domenico Cassini (1625–1712). The student may show that when $a = c$, where $2c = F'F$, the oval becomes a lemniscate; and that when $a < c$ the curve consists of two distinct ovals.

6. In a circle tangent to two parallel lines at the points O and B the chord OR turns about O and cuts the upper tangent at A. If AM is drawn perpendicular to OX and RP perpendicular to AM, the locus of P is as shown in this figure. Find its equation and plot the curve.

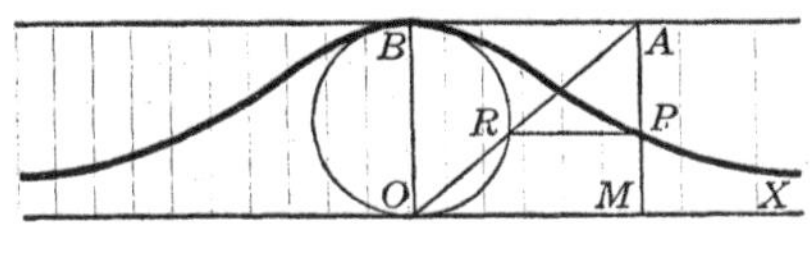

This curve is called the *witch of Agnesi* in honor of Maria Gaetana Agnesi (1718–1799), professor of mathematics at Bologna.

7. When the radius of the rolling circle is half the radius of the fixed circle, the hypocycloid (§ 237) becomes the x axis.

8. A circle of radius r rolls on the inside of a fixed circle of radius $2r$, and a diameter of the small circle is divided by the point P into the segments a, b. Show that the parametric equations of the locus of P are $y = b \sin\theta$ and $x = a \cos\theta$, where θ is the angle indicated in the figure. Eliminate θ from these equations and show that the locus is an ellipse.

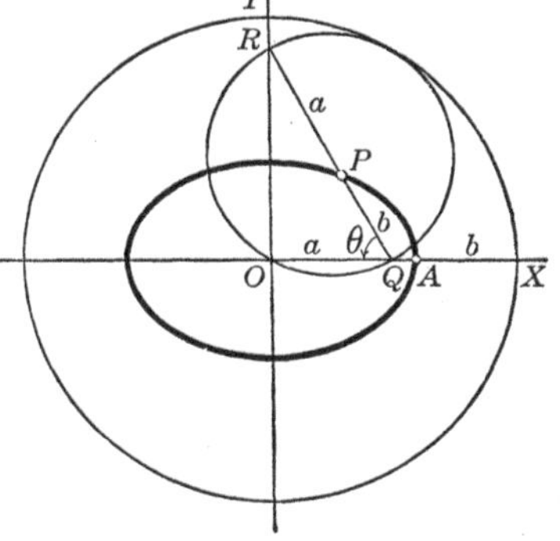

By Ex. 7 the point Q on the smaller circle moves along XO, and the point R on the smaller circle moves along some diameter of the larger circle. Show that R moves along OY. Then when Q moves from X to Q, R moves from O to R, the diameter OX moves to RQ, and the point P moves from A to P.

9. The rectangular equation of the hypocycloid of four cusps is $x^{\frac{2}{3}} + y^{\frac{2}{3}} = (4r)^{\frac{2}{3}}$.

Take $\cos 3\theta = 4\cos^3\theta - 3\cos\theta$ and $\sin 3\theta = 3\sin\theta - 4\sin^3\theta$.

10. Find the parametric equations of the locus of a point P on a circle which rolls on the outside of a fixed circle.

This locus is called an *epicycloid*.

238. Spiral of Archimedes. The graph of the equation $\rho = c\theta$, where c is constant, is called the *spiral of Archimedes.*

Let us plot the graph of the equation $\rho = \frac{1}{2}\theta$. Taking θ as the number of radians in the vectorial angle and the approximate value $3\frac{1}{7}$ for π, the table below shows a few values of ρ.

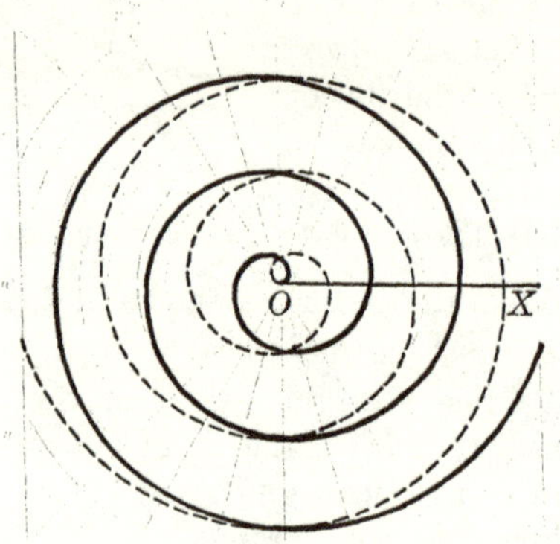

The heavy and dotted lines show the curve for positive and negative values of θ respectively.

This curve is due to Archimedes (287–212 B.C.).

$\theta =$	0	$\frac{1}{6}\pi$	$\frac{1}{3}\pi$	$\frac{1}{2}\pi$	$\frac{2}{3}\pi$	$\frac{5}{6}\pi$	π
$\rho =$	0	0.3	0.5	0.8	1.0	1.3	1.6

239. Logarithmic Spiral. The graph of the equation $\rho = a^{\theta}$ is called the *logarithmic spiral.*

To plot the spiral in the case $\rho = 2^{\theta}$, we may use the form $\log_{10}\rho = \theta \log_{10} 2$ and compute the following values by logarithms:

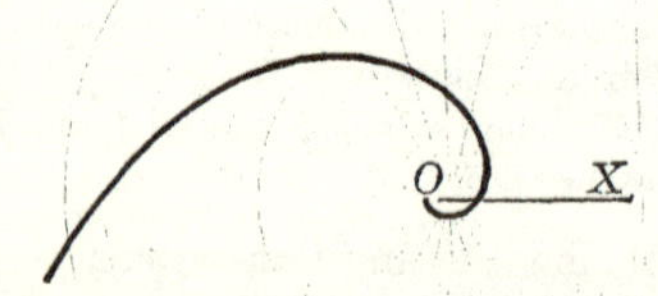

$\theta =$	0	$\frac{1}{6}\pi$	$\frac{1}{3}\pi$	$\frac{1}{2}\pi$	$\frac{2}{3}\pi$	$\frac{5}{6}\pi$	π
$\rho =$	1	1.4	2.1	3.0	4.3	6.1	8.8

The student may also consider the graph for negative values of θ.

240. Exponential Curve. The graph of the equation $y = ca^x$, where a is a positive constant and c is any constant, is called an *exponential curve.*

We shall confine our attention to the curve $y = a^x$, since it is only necessary to multiply its ordinates by c in order to obtain the graph of the equation $y = ca^x$.

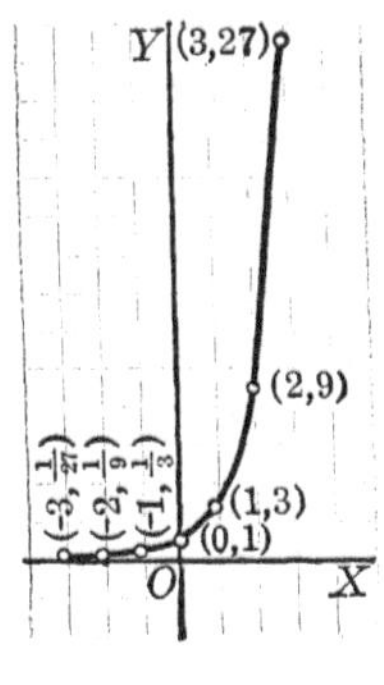

FIG. 1

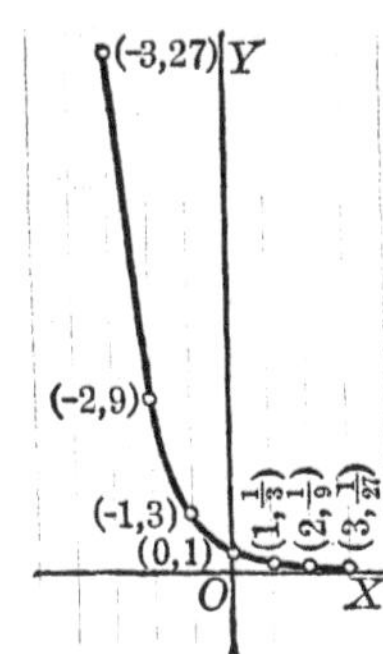

FIG. 2

It is obvious that the curve $y = a^x$ passes through the point (0, 1) and that y is positive for all values of x.

When x is a fraction, say $x = p/q$, $a^x = \sqrt[q]{a^p}$, and we regard this as denoting the positive qth root of a^p. For example, we take $16^{\frac{1}{4}}$ to be 2, not ± 2.

If $a > 1$, a^x increases without limit when x does, and $a^x \rightarrow 0$ when x decreases without limit.

When $a < 1$, $a^x \rightarrow 0$ when x increases without limit, and a^x increases without limit when x decreases without limit.

Considering now the special case $a = 3$, let us plot a few points of the curve $y = 3^x$ and draw the graph. The result is shown in Fig. 1.

Fig. 2 represents the equation $y = (\frac{1}{3})^x$.

In each case points on the graph have been plotted for the values $x = -3, -2, -1, 0, 1, 2, 3$.

241. Logarithmic Curve. The graph of the equation $y = \log_a x$, where the base a is any positive constant, is called a *logarithmic curve.*

Before considering the nature of this curve the student should observe, from the definition of a logarithm, that if

$$x = a^y,$$

then y is the logarithm of x to the base a; that is,

$$y = \log_a x.$$

We shall consider the common case of $a > 1$.

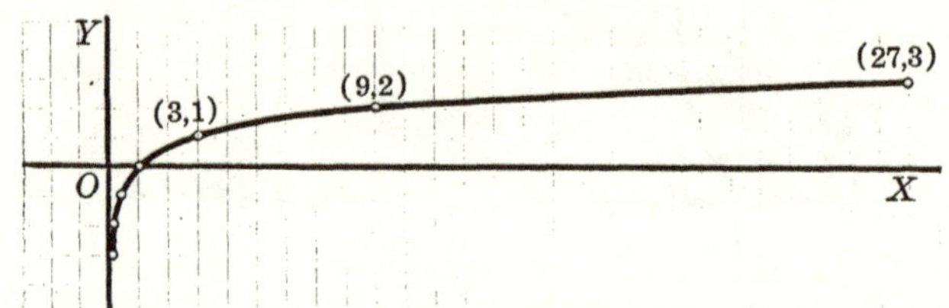

A negative number has no real logarithm since a^k and a^{-k} are both positive, a^{-k} being the same as $1/a^k$. Hence in the equation $y = \log_a x$ it is evident that x must be positive.

When $x < 1$, $\log x$ is negative, as is evident in the cases of $\log_{10} 0.01 = -2$ and $\log_{10} 0.001 = -3$. Furthermore, $\log_a x$ decreases without limit when $x \to 0$, and $\log_a x$ increases without limit when x does.

Consider the case of $a = 3$, and plot the curve $y = \log_3 x$. We may plot the above figure from the following values:

$x =$	$\frac{1}{9}$	$\frac{1}{3}$	1	3	9	27
$y =$	-2	-1	0	1	2	3

The student should draw the graph of $y = \log_a x$ when $a = \frac{1}{3}$.

Jacques Bernoulli had this curve engraved upon his tombstone. The rude figure may still be seen in the cloisters at Basel.

242. Napierian Logarithms. For ordinary computations common logarithms are convenient; that is, logarithms to the base 10. But the student of higher mathematics will meet with various algebraic processes in which a certain other base is far more convenient. This base, called the *Napierian base* in honor of the inventor of logarithms, we shall not now define; but its value is approximately 2.7, and it is usually denoted by e.

The value of e is 2.718281828 $\cdots$. This is not the base used by Napier, but was probably suggested by Oughtred in Edward Wright's translation of Napier's *Descriptio* published in 1618.

243. Relations of Exponential and Logarithmic Curves. The equation $y = \log_a x$ may, according to the definition of a logarithm, be written $x = a^y$. It appears, therefore, that the equations $y = \log_a x$ and $y = a^x$, or $x = a^y$ and $y = a^x$, are related in a simple manner; namely, that if in either equation we interchange x and y, we obtain the other.

When two functions such as $\log_a x$ and a^x are related in this manner, each is called the *inverse* of the other.

The geometric transformation which changes y into x and x into y changes any point $P(x, y)$ to the point $P'(y, x)$. Such a transformation is neither a rotation of the axes through an angle θ in the plane nor a translation of the axes, but is a rotation through the angle 180° of the plane about the line which bisects the first and third quadrants, that is, about the line $y = x$, the axes remaining in the same position as before. These statements may be easily verified by the student.

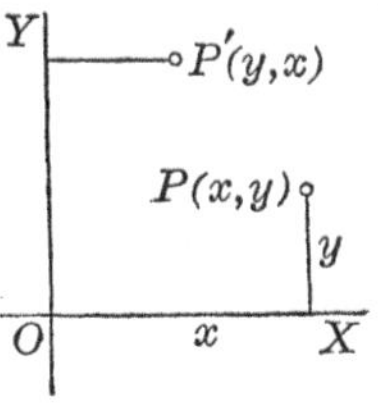

Such a rotation consequently changes the curve $y = \log_a x$ into the curve $y = a^x$, and conversely.

Exercise 60. Exponential and Logarithmic Curves

1. In the equation $y = 2.7^x$, find y when $x = -2, -1, 0, 1, 2, \frac{10}{3}$, and draw the graph.

In some of these cases, as when $x = \frac{10}{3}$, the simplest computation is to take the common logarithm of each member of the equation $y = 2.7^x$. We then have $\log_{10} y = x \log_{10} 2.7$, and for any value of x we may find the value of y from a table of logarithms.

2. Draw the graph of the equation $y = \log_e x$.

Here e is to be taken, as usual, to represent approximately 2.7.

Before assigning values to x, write the equation as $e^y = x$. Then $y \log_{10} e = \log_{10} x$.

3. From the graph of the equation $y = 3^x$, considered as a special case in § 240, draw the graphs of the equations $y = -3^x$ and $y = 2 \times 3^x$.

4. Show that the graph of the equation $y = 3^{-x}$ may be obtained by rotating the graph of the equation $y = 3^x$ about the y axis through an angle of 180°.

5. Draw the graph of the equation $y = e^{2x}$.

6. Draw the graph of the equation $y = \sqrt{e^x}$.

The student should notice that $\sqrt{e^x}$ is the same as $e^{\frac{1}{2}x}$.

7. Draw the graph of the equation $y = 2^{x+1}$ and show that it is the same as the graph of the equation $y = 2 \times 2^x$.

8. Draw the graph of the equation $y = e^{-x^2}$.

This graph is known as the *probability curve.*

9. At 4% interest, compounded annually, the amount of $1 at the end of x years is 1.04^x dollars. Draw a graph showing the growth of the amount for 10 yr.

10. Draw the graph of the equation $y = \log_{10} x$.

11. Draw the graph of the equation $y = \log_{10} 2x$.

It should be recalled that $\log 2x = \log x + \log 2$. Since the graph of $y = \log_{10} x$ was drawn in Ex. 10, the graph of $y = \log_{10} 2x$ may be found by adding $\log_{10} 2$, or about 0.3, to each ordinate of the graph of $y = \log_{10} x$; that is, the entire graph of $y = \log_{10} x$ is moved up about 0.3.

244. Trigonometric Curves. In § 14 we drew the graph of the equation $y = \sin x$, and the student should now find it easy to draw the graphs of the equations $y = \cos x$, $y = \tan x$, $y = \cot x$, $y = \sec x$, $y = \csc x$. It is advisable, however, to become so familiar with the graphs of $y = \sin x$, $y = \cos x$, and $y = \tan x$ that they may be sketched at once without stopping for computations, since these three curves, and particularly the first two, are of considerable importance in many connections.

In the case of $y = \tan x$ it may be noted that the vertical line drawn at $x = 90°$ is an asymptote to the graph. This is evident for the reason that $\tan x$ increases without limit when $x \longrightarrow 90°$ from the left, and $\tan x$ decreases without limit when $x \longrightarrow 90°$ from the right.

Similar remarks apply to the cases of $y = \cot x$, $y = \sec x$, and $y = \csc x$. Any convenient units may be chosen for x and y. When x is given in radians, use any convenient length on OX for π radians.

245. Inverse Trigonometric Curves. The graphs of the equations $y = \sin^{-1} x$, $y = \cos^{-1} x$, and so on are called *inverse trigonometric curves.*

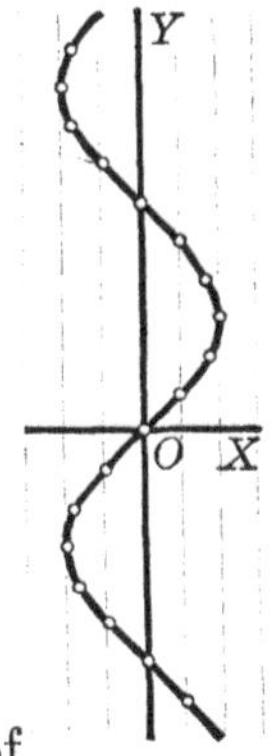

The symbol $\sin^{-1} x$, also written arc sin x, means an angle whose sine is x, and hence the equation $y = \sin^{-1} x$ may be written $x = \sin y$.

We shall consider at this time only the equation $y = \sin^{-1} x$, and shall mention two simple methods of drawing the graph.

As a first method we may write the equation $y = \sin^{-1} x$ in the form $x = \sin y$, assign values to y, and then compute the corresponding values of x, thus obtaining points on the graph.

As a second method, knowing the graph of $y = \sin x$, we may obtain from it the graph of $x = \sin y$ by rotating through the angle 180° the curve $y = \sin x$ about the line $y = x$, as explained in § 243.

Exercise 61. Trigonometric Curves

1. Draw the graph of the equation $y = \sin x$.

In the problems on this page, x represents radians. As convenient units, the student may take a length of $\frac{1}{4}$ in. to represent an angle of $\frac{1}{6}\pi$ radians on the x axis, and $\frac{1}{2}$ in. to represent 1 on the y axis.

2. Draw the graph of the equation $y = \cos x$.

3. Show that in the interval $-\frac{1}{2}\pi < x < \frac{1}{2}\pi$ the graph of the equation $y = \tan x$ passes through the points $(-\frac{1}{4}\pi, -1)$, $(0, 0)$, $(\frac{1}{4}\pi, 1)$, and approaches from above the vertical asymptote $x = -\frac{1}{2}\pi$ and from below the vertical asymptote $x = \frac{1}{2}\pi$. Draw the graph in this interval, and draw and discuss the graph in an interval to the right of this one.

4. Draw and discuss the graph of the equation $y = \cot x$ in the interval from $x = 0$ to $x = \pi$, and in one more interval.

5. Draw and discuss the graph of the equation $y = \sec x$ in the intervals $\frac{1}{2}\pi < x < \frac{3}{2}\pi$ and $\frac{3}{2}\pi < x < \frac{5}{2}\pi$.

6. Draw the graph of the equation $y = \sin 2x$ and compare its period with that of $y = \sin x$.

It is evident from the graph of the equation $y = \sin x$ (§ 14) that the sine returns periodically to any given value, and that this *period* is 2π.

7. Draw the graph of the equation $y = \cos \frac{1}{2} x$ and compare its period with that of $y = \cos x$.

8. The translation of axes in which the origin is changed to the point $(\frac{1}{2}\pi, 0)$ transforms the equation $y = \sin x$ into the equation $y = \cos x$.

9. Show how the graph of the equation $y = \cot x$ may be obtained from that of $y = \tan x$ by a translation of axes.

10. Show how the graph of the equation $y = \sin(x + a)$ may be obtained from that of $y = \sin x$ by a translation of axes.

11. Draw the graph of the equation $y = \cos(x + \frac{1}{6}\pi)$.

12. Draw the graph of the equation $y = \sin(x - \frac{1}{3}\pi)$.

13. Draw the graph of the equation $y = \cos(3x - 2\pi)$.

Exercise 62. Review

1. Find the equation of the locus of the foot of the perpendicular from the point $(-a, 0)$ upon the tangent to the parabola $y^2 = 4\,ax$, and draw the locus.

This locus is called a *strophoid*, which means "like a twisted band."

2. The locus of the foot of the perpendicular from the vertex of the parabola $y^2 = -4\,ax$ upon the tangent is a cissoid.

3. If r denotes the radius of a circle rolling on a straight line, b the distance of any point on the radius from the center of the circle, and θ the same angle as in § 234, show that the equations of the locus of the generating point on the radius are $x = r\theta - b \sin \theta$ and $y = r - b \cos \theta$.

This locus is called a *trochoid*, which means "wheel-like."

4. Draw the curve in Ex. 3 when $b < r$ and when $b > r$.

When $b < r$, the curve is called the *prolate cycloid*, and when $b > r$, it is called the *curtate cycloid*. When $b = r$, it is evident that the curve is the cycloid, as described in § 234.

5. Assuming that c in the equation $\rho^2 = c/\theta$ is a positive constant, and taking the value $\rho = \sqrt{c/2\,\pi}$ when $\theta = 2\,\pi$, show that ρ increases without limit when θ decreases from $2\,\pi$ to 0, and that $\rho \rightarrow 0$ when θ increases from $2\,\pi$ without limit. Draw the curve, and also consider the case of $\rho = -\sqrt{c/2\,\pi}$.

This curve is called the *lituus*, which means "a bishop's staff." The reason for the name will be evident from the figure.

6. Draw the graph of the equation $\rho = c/\theta$.

This curve is called the *reciprocal spiral*.

7. Let $P(\rho, \theta)$ be any point on the reciprocal spiral $\rho = c/\theta$. Draw a circular arc clockwise from P and meeting the polar axis at the point Q, and prove that the arc PQ is constant.

8. If $P_1(\rho_1, \theta_1)$, $P_2(\rho_2, \theta_2)$, $P_3(\rho_3, \theta_3)$ are any three points of the logarithmic spiral $\rho = a^\theta$, and if the vectorial angles $\theta_1, \theta_2, \theta_3$ are in arithmetic progression, then the radius vectors ρ_1, ρ_2, ρ_3 are in geometric progression.

9. Show that the spiral of Archimedes is cut by any line through the origin in an infinite number of points. Denote by P and Q two such successive points, and prove that the length PQ is constant when the line revolves about the origin.

If P is the point (ρ_1, θ_1), then Q is $(\rho_2, \theta_1 + 2\pi)$.

10. This figure represents a wheel rotating about the fixed center C and driven by the rod AQ, the point A sliding on the line AC. Show that $x = (b - a)\cos\theta \pm \sqrt{r^2 - a^2\sin^2\theta}$ and $y = b\sin\theta$, where $a = AQ$, $b = AP$, and $\theta = \angle CAQ$, are the equations of the locus of a point P on AQ.

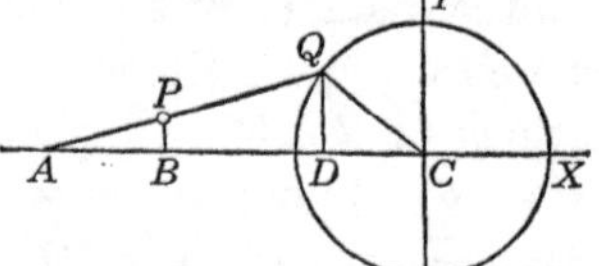

Draw QD perpendicular to AC. Then $DQ = a\sin\theta$, and x can be found from $x = CB = -BC = -(AD - AB + DC)$.

11. Draw the locus in Ex. 10 when $a = 12$, $b = 5$, $r = 8$.

12. The locus in Ex. 10 is an ellipse when $a = r$.

13. In the figure of Ex. 10 draw a line perpendicular to AC at A, and let it meet CQ produced at R. Find the parametric equations of the locus of R and draw the locus. Prove that when $a = r$ the locus is a circle.

Take as parameter either the angle θ or the angle ACQ, denoting the latter angle by ϕ. This is an important locus, R being known as the *instantaneous center* of the motion of AQ.

14. If the circle $x^2 + y^2 = r^2$ is represented by parametric equations, one of which is $x = t - r$, find the equation which represents y in terms of t.

15. A line OA begins to turn about the point O at the rate of 2° per minute while a point P begins to move from O along OA at the rate of 2 in. per minute. Find the parametric equations of the locus of P.

16. Find the parametric equations of the locus of the end of a cord of negligible diameter unwinding from a fixed circle.

17. Draw the graph of the equation $y = \sin 2x$ and obtain the graph of $y = \sin(2x - \frac{1}{4}\pi)$ by a translation of axes.

CHAPTER XIII

POINT, PLANE, AND LINE

246. Locating a Point. The position of a point in space may be determined by its distances from three fixed planes that meet in a point. The three fixed planes are called the *coordinate planes*, their three lines of intersection are called the *coordinate axes*, and their common point is called the *origin*. In the work that follows we shall employ coordinate planes that intersect each other at right angles, these being the ones most frequently used.

Let XOY, YOZ, and ZOX be three planes of indefinite extent intersecting each other at right angles in the lines XX', YY', and ZZ'. These coordinate planes are called the xy plane, the yz plane, and the zx plane respectively, and the axes XX', YY', ZZ' are called the axes of x, y, z respectively.

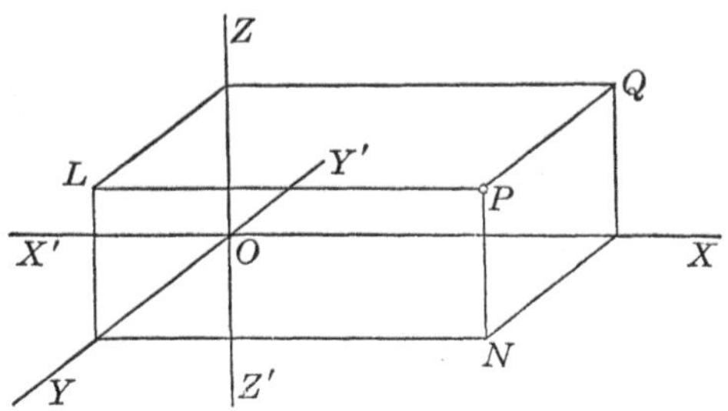

The distances LP, QP, NP of a point P in space from the coordinate planes are called the *coordinates* x, y, z of P, and the point is denoted by (x, y, z) or by $P(x, y, z)$.

There are various other methods of locating a point in space besides the one described above. Two of the most important of these methods are explained in § 282.

247. Convention of Signs. If through any point P in space we take the three planes parallel to the coordinate planes, we have with the coordinate planes a rectangular parallelepiped, as shown in the left-hand figure below.

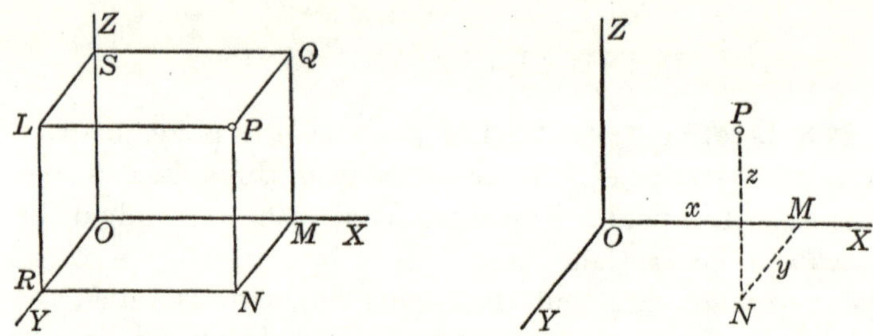

The coordinate x of P is LP, SQ, RN, or OM; the coordinate y of P is QP, SL, MN, or OR; and the coordinate z of P is NP, RL, MQ, or OS.

One of the most convenient ways of drawing the coordinates of a point is shown in the figure at the right, y being drawn from the end of x, and z being drawn from the end of y.

Choosing the directions OX, OY, OZ as the positive directions along the axes, we regard the coordinate x of a point as positive when it has the direction of OX, and as negative when it has the direction of OX'; we regard the coordinate y of a point as positive when it has the direction of OY and as negative when it has the direction of OY'; and similarly for the coordinate z.

The coordinate planes divide space into eight portions, called *octants*. The octant XYZ, in which the coordinates x, y, z of a point are all positive, is called the *first octant*.

It is not necessary to assign to the other seven octants such a specified order as second, third, and so on.

For the sake of simplicity we shall, when the conditions allow, draw the diagrams in the first octant.

PROBLEM. DISTANCE BETWEEN TWO POINTS

248. *To find the distance between two points* P_1 (x_1, y_1, z_1) *and* P_2 (x_2, y_2, z_2).

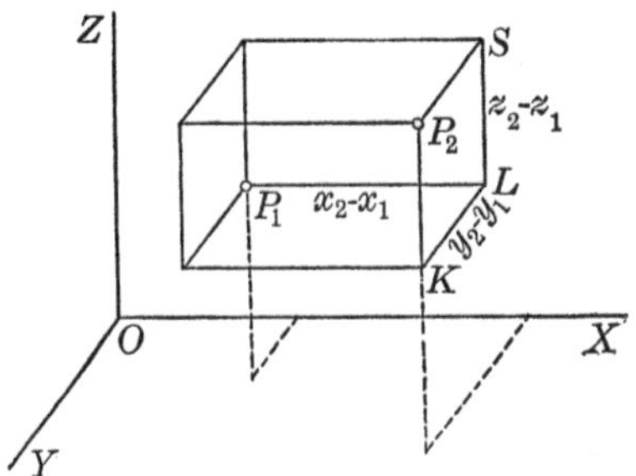

Solution. Through each of the two given points P_1 and P_2 pass planes parallel to the three coordinate planes, thus forming the rectangular parallelepiped whose diagonal is P_1P_2, and whose edges P_1L, LK, KP_2 are parallel to the axes of x, y, z respectively.

Then $$\overline{P_1P_2}^2 = \overline{P_1L}^2 + \overline{LK}^2 + \overline{KP_2}^2.$$

Now P_1L is the difference between the distances of P_1 and P_2 from the yz plane, so that $P_1L = x_2 - x_1$. For a like reason $LK = y_2 - y_1$, and $LS = KP_2 = z_2 - z_1$. Hence, denoting the distance P_1P_2 by D and substituting in the above equation, we have

$$D = \sqrt{(x_2 - x_1)^2 + (y_2 - y_1)^2 + (z_2 - z_1)^2},$$

which is the required formula.

If P_1 and P_2 are taken in any other octant than the first, it will be found that $P_1L = x_2 - x_1$, $LK = y_2 - y_1$, and $KP_2 = z_2 - z_1$.

249. COROLLARY. *The distance from the origin to any point* $P(x, y, z)$ *is given by the formula*

$$\overline{OP}^2 = x^2 + y^2 + z^2.$$

250. Direction Angles of a Line. Denote by α, β, γ the angles which a directed line makes with the positive directions of the axes of x, y, z respectively. These angles are called the *direction angles* of the line, and their cosines are called the *direction cosines* of the line.

The angles α, β, γ are always positive and never greater than 180°. They correspond to one direction along the line, while to the opposite direction correspond the angles $180° - \alpha$, $180° - \beta$, $180° - \gamma$, the cosines of which are equal to $-\cos\alpha$, $-\cos\beta$, $-\cos\gamma$.

THEOREM. DIRECTION COSINES OF A LINE

251. *The sum of the squares of the direction cosines of any line is equal to 1.*

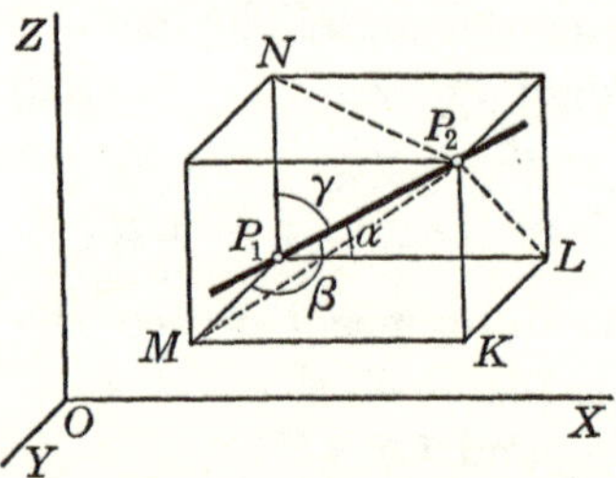

Proof. Through P_1 and P_2, any two given points, pass planes parallel to the coordinate planes, thus forming the rectangular parallelepiped whose diagonal is P_1P_2. Then from the right triangles P_1LP_2, P_1MP_2, P_1NP_2 we have

$$\cos\alpha = \frac{x_2 - x_1}{D}, \quad \cos\beta = \frac{y_2 - y_1}{D}, \quad \cos\gamma = \frac{z_2 - z_1}{D},$$

where $D = P_1P_2$. Squaring and adding, we have (§ 248)

$$\cos^2\alpha + \cos^2\beta + \cos^2\gamma = 1.$$

The above formulas for $\cos\alpha$, $\cos\beta$, $\cos\gamma$ show how to find the direction cosines of the line through two given points.

252. Corollary 1. *If ρ is the distance from the origin to any point $P(x, y, z)$, then*

$$x = \rho \cos\alpha, \quad y = \rho \cos\beta, \quad z = \rho \cos\gamma.$$

253. Corollary 2. *A line can be found having direction cosines proportional to any three numbers a, b, c.*

Using a, b, c as coordinates, locate the point $P(a, b, c)$, and from the origin O draw OP. We then have $OP = \sqrt{a^2 + b^2 + c^2}$, and the direction cosines of OP, or of any line parallel to OP, are (§ 251)

$$\cos\alpha = \frac{a}{\pm\sqrt{a^2 + b^2 + c^2}},$$

$$\cos\beta = \frac{b}{\pm\sqrt{a^2 + b^2 + c^2}},$$

$$\cos\gamma = \frac{c}{\pm\sqrt{a^2 + b^2 + c^2}},$$

and these values are easily seen to be proportional to a, b, c.

The plus sign for the radical corresponds to one direction along the line, and the minus sign corresponds to the opposite direction. Whenever we are not concerned with the directions along the line we shall take the value given by the plus sign.

Exercise 63. Points and Direction Cosines

1. Locate the points $(6, 2, 5)$, $(-4, 1, 7)$, $(2, -1, -4)$, $(0, 6, -3)$, $(0, 0, 2)$, $(0, 2, 0)$.

2. Show that each of the points $(3, 1, 0)$, $(-2, 0, -3)$, $(0, 4, -1)$, $(a, b, 0)$, $(0, b, 0)$ is in one of the coördinate planes.

3. Show that each of the points $(4, 0, 0)$, $(0, 7, 0)$, $(0, 0, -5)$, $(0, r, 0)$ is on one of the coordinate axes.

4. Show that the numbers 0.2, -1, 0.4 cannot be the three direction cosines of a line.

5. Show that the numbers $\frac{1}{2}$, $\frac{1}{3}$, $\frac{2}{5}$ cannot be the three direction cosines of a line.

6. If two direction cosines of a line are $\frac{1}{2}$, $\frac{1}{3}$, find the other.

Problem. Division of a Line

254. *To find the coordinates of the point which divides the line joining the points P_1 (x_1, y_1, z_1) and P_2 (x_2, y_2, z_2) in a given ratio $m_1 : m_2$.*

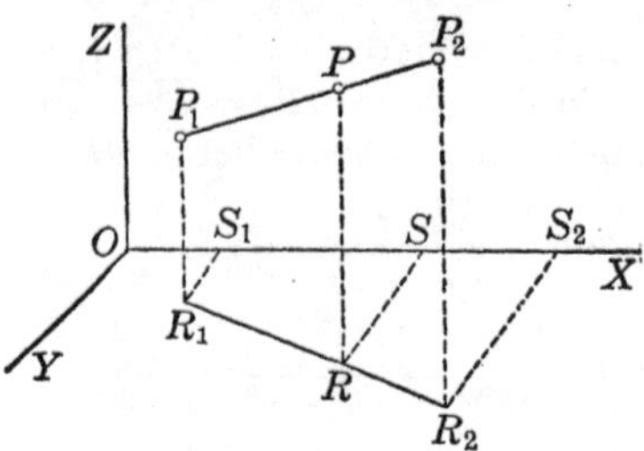

Solution. Denote by P the required point and draw its coordinates x, y, z. Draw the coordinates x_1, y_1, z_1 and x_2, y_2, z_2 of the given points P_1 and P_2 respectively, as shown in the figure.

Then
$$\frac{P_1P}{PP_2} = \frac{R_1R}{RR_2} = \frac{S_1S}{SS_2},$$
or
$$\frac{m_1}{m_2} = \frac{x - x_1}{x_2 - x}.$$

Solving for x, we have
$$x = \frac{m_2x_1 + m_1x_2}{m_1 + m_2}.$$

By the aid of other figures analogous to the above, the student should show that
$$y = \frac{m_2y_1 + m_1y_2}{m_1 + m_2},$$
and
$$z = \frac{m_2z_1 + m_1z_2}{m_1 + m_2}.$$

The above method and the results also hold for oblique axes.

Exercise 64. Direction Cosines and Points of Division

Find the length and the direction cosines of the line joining each of the following pairs of points:

1. (4, 2, – 2), (6, – 4, 1).
2. (3, 2, 4), (– 3, 5, 5).
3. (2, – 4, –1), (2, 7, 6).
4. (5, 0, 0), (0, 4, – 2).

5. Locate the point $P(2, 3, 6)$ and draw PK perpendicular to the xy plane and KL perpendicular to OX. Draw PL and find $\cos\alpha$ for the line OP. Draw other lines and in the same way find $\cos\beta$ and $\cos\gamma$.

6. Locate the points $P_1(4, 2, 4)$ and $P_2(8, 5, 10)$. Draw the parallelepiped having P_1P_2 as diagonal and having edges parallel to the axes. Find the length of each edge, the length of P_1P_2, and the direction cosines of P_1P_2.

Find the direction cosines of each of the following lines:

7. The line in the yz plane bisecting the angle YOZ.

8. The x axis; the y axis; the z axis.

9. The lines for which $\alpha = 60°$ and $\beta = 120°$.

10. The line for which $\alpha = \beta = \gamma$.

11. Find the direction cosines of a line, given that they are proportional to 12, – 4, 6; to – 3, – 2, 3.

12. The line that passes through the points (– 6, 6, 2) and (– 6, 2, – 2) has the direction angles 90°, 45°, 45°.

13. The coordinates of the mid point of the line which joins the points $P_1(x_1, y_1, z_1)$ and $P_2(x_2, y_2, z_2)$ are $x = \frac{1}{2}(x_1 + x_2)$, $y = \frac{1}{2}(y_1 + y_2)$, $z = \frac{1}{2}(z_1 + z_2)$.

14. Find the coordinates of the point which divides in the ratio 5 : 2 the line that joins the points $P_1(-2, -3, -1)$ and $P_2(-5, -2, 4)$.

15. The equation $x^2 + y^2 + z^2 = 25$ is true for all points (x, y, z) on the sphere with radius 5 and with center at the origin, and for no other points.

16. The four lines joining the vertices of a tetrahedron to the points of intersection of the medians of the opposite faces meet in a point R which divides each line in the ratio $3:1$.

The point R is called the *center of mass* of the tetrahedron.

Oblique axes may be employed, the axes lying along three concurrent edges of the tetrahedron.

17. If from the mid point of each edge of the tetrahedron in Ex. 16 a line is drawn to the mid point of the opposite edge, the four lines thus drawn meet in a point which bisects each of the lines.

18. If the point (x, y, z) is any point on the sphere with center (a, b, c) and radius r, show that $(x-a)^2+(y-b)^2+(z-c)^2=r^2$.

Observe that this equation of the sphere is similar to that of the circle.

19. Where are all points (x, y, z) for which $x = 6$? $x = 0$?

20. Find the locus of all points for which $z = 2$; for which $y = 4$; for which $z = 2$ and $y = 4$ at the same time.

21. Find an equation which is true for all points (x, y, z) which are equidistant from the points $(2, 3, -7)$ and $(-1, 3, 3)$.

22. Through the point $A(3, 2, 8)$ draw a line parallel to OZ. If P is a point on this line such that $OP = 7$, find the coordinates of P.

23. If the point $P(x, y, z)$ moves so that its distance from the point $A(2, 3, -3)$ is equal to its distance from the xy plane, find an equation in x, y, z which is true for all positions of P.

24. Find the coordinates of a point in the xy plane which is 5 units from the origin and is equidistant from the points $A(3, 4, 2)$ and $B(5, 1, 1)$.

25. Find a point that is equidistant from the three points $A(2, 0, 0)$, $B(1, 2, -1)$, $C(2, -1, 0)$.

The number of such points is, of course, unlimited.

26. Find the center and the radius of the sphere circumscribed about the tetrahedron having the vertices $A(2, 0, 0)$, $B(1, 2, -1)$, $C(2, -1, 0)$, and $O(0, 0, 0)$.

255. Projection of a Line Segment. Let P_1P_2 in the figure of § 256 be any line segment, and let l be any straight line in space. Pass planes through P_1 and P_2 perpendicular to l at the points R_1 and R_2 respectively.

The points R_1 and R_2 are called the *projections of the points P_1 and P_2 on l*, and the segment R_1R_2 is called the *projection of the line segment P_1P_2 on l.* If the segment is P_2P_1, taken in the opposite direction, its projection is R_2R_1.

There are other kinds of projection, but we shall employ only the projection described above, which is called *orthogonal projection.*

PROBLEM. LENGTH OF PROJECTION

256. *To find the length of the projection of a given line segment on a given line.*

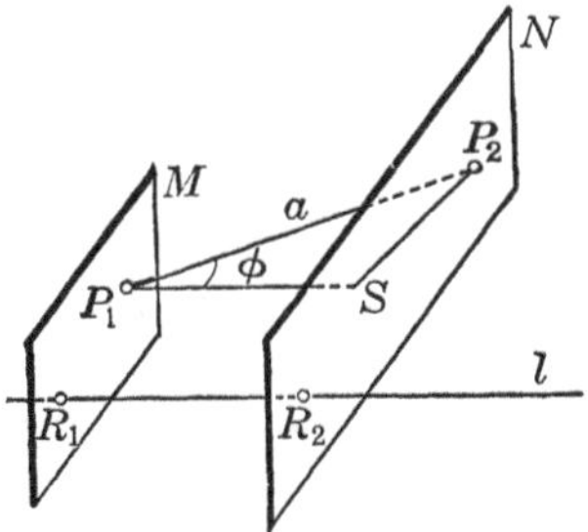

Solution. Let the given segment be P_1P_2, or a; ϕ the angle which P_1P_2 makes with the given line l; and M and N planes perpendicular to l through P_1 and P_2 respectively.

Through P_1 draw a line parallel to l, meeting the plane N at the point S, and draw P_2S. Then the angle $SP_1P_2 = \phi$, and P_1S is equal to the projection R_1R_2 which we are to find.

But obviously $P_1S = a \cos \phi$, and hence the length h of the projection of a on l may be found from the formula

$$h = a \cos \phi.$$

THEOREM. PROJECTION OF A BROKEN LINE

257. *The sum of the projections of the parts AB, BC, CD of a broken line ABCD on any line l is equal to the projection of the straight line AD on l.*

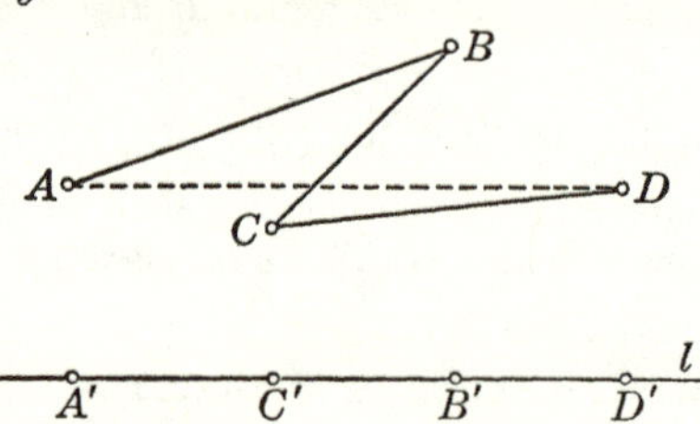

Proof. Let the projections of A, B, C, D on l be A', B', C', D'. Then the projections of the segments AB, BC, CD, AD on l are $A'B'$, $B'C'$, $C'D'$, $A'D'$ respectively, and the proof of the theorem follows from the fact that

$$A'B' + B'C' + C'D' = A'D'.$$

It should be remembered that the projections are directed segments; thus, if $A'B'$ is positive, $B'C'$ is negative. The theorem is obviously true for a broken line having any number of parts. It is not assumed that the parts of the broken line all lie in one plane.

258. Equation of a Surface. If $P(x, y, z)$ is any point on the sphere of radius r having the origin as center, then $x^2 + y^2 + z^2 = r^2$, since this equation expresses the fact that $\overline{OP}^2 = r^2$. The equation $x^2 + y^2 + z^2 = r^2$ is true for all points (x, y, z) on the sphere, and for no other points, and is called the *equation of the sphere.*

In general, the *equation of a surface* is an equation which is satisfied by the coordinates of every point on the surface, and by the coordinates of no other points.

The subsequent work is concerned almost entirely with equations of the first and second degrees in x, y, and z, and with the surfaces which these equations represent.

Problem. Normal Equation of a Plane

259. *To find the equation of a plane when the length of the perpendicular from the origin and the direction cosines of this perpendicular are given.*

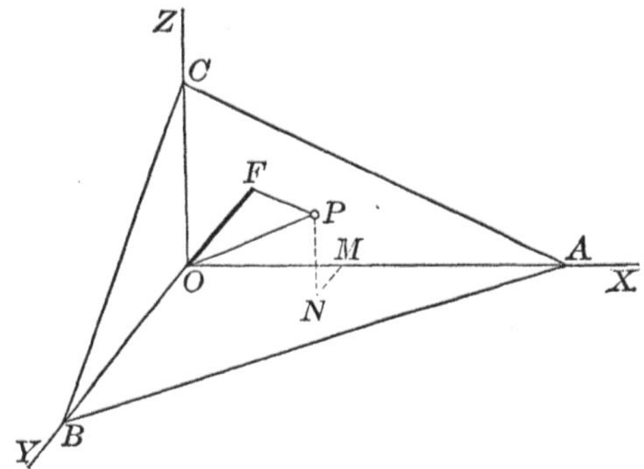

Solution. Letting OF be the perpendicular to the plane ABC from the origin O, denote the length of OF by p and the direction angles by α, β, γ. Let P be any point in the plane, let OM, MN, NP be its coordinates x, y, z, and draw OP. Then the projection of OP on OF is equal to the sum of the projections of OM, MN, NP on OF (§ 257). But as the plane is perpendicular to OF, p is the projection of OP on OF, and the projections of OM, MN, NP on OF are respectively $x \cos \alpha$, $y \cos \beta$, $z \cos \gamma$ (§ 256).

Hence we have the required equation

$$\boldsymbol{x \cos \alpha + y \cos \beta + z \cos \gamma = p.}$$

This equation is called the *normal equation* of the plane, and in work with planes it is of great importance.

For example, if the direction cosines of the line OF in the figure above are $\frac{3}{7}$, $\frac{6}{7}$, $\frac{2}{7}$, and if $OF = 5$, the equation of the plane perpendicular to OF at F is $\frac{3}{7}x + \frac{6}{7}y + \frac{2}{7}z = 5$, which may be written $3x + 6y + 2z = 35$.

THEOREM. EQUATION OF THE FIRST DEGREE

260. *Every equation of the first degree in x, y, and z represents a plane.*

Proof. The general equation of the first degree in x, y, z is

$$Ax + By + Cz + D = 0. \qquad (1)$$

Dividing by $\sqrt{A^2 + B^2 + C^2}$, denoted by S when preceded by the sign opposite to that of D, we have

$$\frac{A}{S}x + \frac{B}{S}y + \frac{C}{S}z = -\frac{D}{S}. \qquad (2)$$

But these coefficients of x, y, z are the direction cosines of some line l. § 253

Hence (2) is the equation of the plane perpendicular to l at the distance $-D/S$ from the origin. § 259

261. COROLLARY 1. *To reduce any equation of the first degree in x, y, z to the normal form, write it in the form $Ax + By + Cz + D = 0$ and divide by $\sqrt{A^2 + B^2 + C^2}$ preceded by the sign opposite to that of D.*

For example, to reduce the equation $3x - 6y + 2z + 14 = 0$, we divide by $-\sqrt{9 + 36 + 4}$, or -7, obtaining $-\frac{3}{7}x + \frac{6}{7}y - \frac{2}{7}z = 2$. Thus the length of the perpendicular from the origin to the plane is 2, and the direction cosines of the perpendicular are $-\frac{3}{7}, \frac{6}{7}, -\frac{2}{7}$.

262. COROLLARY 2. *The direction cosines of any straight line perpendicular to the plane $Ax + By + Cz + D = 0$ are*

$$\frac{A}{\pm\sqrt{A^2+B^2+C^2}}, \quad \frac{B}{\pm\sqrt{A^2+B^2+C^2}}, \quad \frac{C}{\pm\sqrt{A^2+B^2+C^2}},$$

and are proportional to A, B, C.

The notation $a : b : c = d : e : f$ will be used to indicate that a, b, c are proportional to d, e, f; in other words, that the ratios of a, b, and c are equal to the ratios of d, e, and f respectively; or, again, that $a : d = b : e = c : f$.

PROBLEM. INTERCEPT EQUATION OF A PLANE

263. *To find the equation of a plane in terms of the intercepts of the plane on the axes.*

Solution. Denote the intercepts of a given plane on the axes of x, y, z by a, b, c respectively. Let the equation

$$Ax + By + Cz = D$$

represent the plane. This equation must be satisfied by the coordinates of the points $(a, 0, 0)$, $(0, b, 0)$, $(0, 0, c)$, the end points of the intercepts.

Hence $\qquad Aa = D,\ Bb = D,\ Cc = D;$

that is, $\qquad a = D/A,\ b = D/B,\ c = D/C.$

But the equation of the plane may be written in the form

$$\frac{x}{D/A} + \frac{y}{D/B} + \frac{z}{D/C} = 1,$$

and hence $\qquad \boldsymbol{\frac{x}{a} + \frac{y}{b} + \frac{z}{c} = 1.}$

264. Plane through Three Points. The equation of the plane $Ax + By + Cz = D$ has four terms. If we divide by D, we have the form $lx + my + nz = 1$. Hence the equation involves three essential constants, and a plane is determined by any three independent and consistent conditions.

For example, find the equation of the plane which passes through the three points $P(1, -2, 2)$, $Q(6, 2, 4)$, and $R(4, 1, 5)$.

Let the equation $lx + my + nz = 1$ represent the plane. Since P, Q, and R are on the plane, their coordinates must satisfy the equation of the plane. Substituting, we have

$$l - 2m + 2n = 1, \quad 6l + 2m + 4n = 1, \quad 4l + m + 5n = 1.$$

Solving, we find that $l = \frac{1}{5}$, $m = -\frac{3}{10}$, and $n = \frac{1}{10}$. Therefore the required equation is $\frac{1}{5}x - \frac{3}{10}y + \frac{1}{10}z = 1$, or $2x - 3y + z = 10$.

This method fails if the plane contains the origin, for the equation is then $Ax + By + Cz = 0$.

PROBLEM. PLANE PERPENDICULAR TO A LINE

265. *To find the equation of the plane passing through the point $P_1(x_1, y_1, z_1)$ and perpendicular to a line l whose direction cosines are known.*

Solution. Represent the plane by the equation

$$Ax + By + Cz = D. \tag{1}$$

Since the plane is perpendicular to the line l, the coefficients A, B, C must be chosen proportional to the known direction cosines of l. § 262

Since the plane passes through $P_1(x_1, y_1, z_1)$, we have

$$Ax_1 + By_1 + Cz_1 = D. \tag{2}$$

Putting this value of D in (1), we have the equation

$$\boldsymbol{A(x - x_1) + B(y - y_1) + C(z - z_1) = 0,}$$

where A, B, C are any three numbers proportional to the direction cosines of the given line l.

Exercise 65. Equations of Planes

Given the points $A(2, 1, -1)$, $B(-2, 2, 4)$, $C(3, 2, -2)$, find in each case the equation of the plane which:

1. Passes through A, B, and C.

2. Passes through A perpendicular to BC.

3. Passes through B perpendicular to AC.

4. Is perpendicular to AB at A.

5. Is perpendicular to AC at the mid point of AC.

6. Passes through A, making equal intercepts on the axes.

7. Passes through B and C parallel to the z axis.

8. Passes through A and B, the y intercept exceeding the z intercept by 5.

9. Contains the point (3, 1, 0) and the line AB.

THEOREM. PARALLEL PLANES

266. *If* $Ax + By + Cz = D$ *and* $A'x + B'y + C'z = D'$ *represent two parallel planes,* $A : B : C = A' : B' : C'$, *and conversely.*

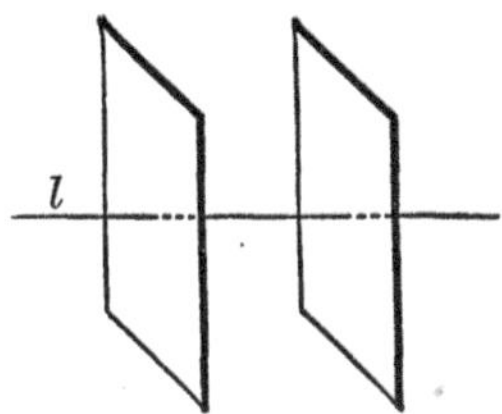

Proof. Let l denote a line perpendicular to the first plane, and let d, e, f be three numbers proportional to the direction cosines of l.

Then $$A : B : C = d : e : f. \qquad \S\ 262$$

Since the planes are parallel, l is perpendicular to the second plane also.

Then $$A' : B' : C' = d : e : f,$$

and hence $$A : B : C = A' : B' : C'.$$

Conversely, if $A : B : C = A' : B' : C'$, we have

$$A : B : C = d : e : f,$$

and $$A' : B' : C' = d : e : f.$$

Hence the second plane is perpendicular to l, and the planes are parallel.

267. COROLLARY. *The equation of the plane that passes through the point* $P_1(x_1, y_1, z_1)$ *and is parallel to the plane* $Ax + By + Cz = D$ *is*

$$A(x - x_1) + B(y - y_1) + C(z - z_1) = 0.$$

The desired equation represents a plane parallel to the given plane $Ax + By + Cz = D$ (§ 266), and since the equation is satisfied by the coordinates of $P_1(x_1, y_1, z_1)$, the plane also passes through P_1.

AG

268. Special Planes. It is important to find what planes are represented by such special equations of the first degree as $x = a$ and $Ax + By = D$. It is easily seen that

The equation $x = a$ represents the plane parallel to the yz plane at the distance a.

For all points (x, y, z) for which x has the same value a lie in that plane, and all points in that plane have the same value of x; that is, $x = a$.

Similarly, *the equation $y = b$ represents the plane parallel to the zx plane at the distance b, and $z = c$ represents the plane parallel to the xy plane at the distance c.*

Consider the plane $Ax + By = D$, denoting by α, β, γ the direction angles of a line l perpendicular to the plane.

Then we see that $\cos\gamma = 0$ (§ 262). Hence the line l is perpendicular to the z axis, and the plane, being perpendicular to l, is parallel to the z axis. Therefore

The plane $Ax + By = D$ is parallel to the z axis, the plane $Ax + Cz = D$ is parallel to the y axis, and the plane $By + Cz = D$ is parallel to the x axis.

A single illustration will make the above statements clearer. For example, find the equation of the plane through $A(3, 1, -1)$, $B(2, -1, 3)$ and parallel to OX.

Let the equation of the plane be

$$By + Cz = D. \tag{1}$$

Dividing by D, we may write this equation in the form

$$ly + mz = 1, \tag{2}$$

where $l = B/D$ and $m = C/D$.

Since A and B are on the plane, we have from (2)

$$l - m = 1,$$

and

$$-l + 3m = 1.$$

Solving for l and m, we find that $l = 2$, $m = 1$, and hence the required equation is

$$2y + z = 1.$$

PROBLEM. DISTANCE OF A POINT FROM A PLANE

269. *To find the perpendicular distance of a given point from a given plane.*

Solution. Let the normal equation of the given plane be

$$x \cos \alpha + y \cos \beta + z \cos \gamma = p. \qquad \S\, 259$$

Let $P_1(x_1, y_1, z_1)$ be the given point, and let the plane M pass through P_1 parallel to the given plane. Denote by d the distance from the given plane to this second plane. Then d is the distance we are to find.

The equation of the plane M is evidently

$$x \cos \alpha + y \cos \beta + z \cos \gamma = p + d;$$

and since it is given that this plane passes through the point $P_1(x_1, y_1, z_1)$, we have

$$x_1 \cos \alpha + y_1 \cos \beta + z_1 \cos \gamma = p + d;$$

whence $$\mathbf{d = x_1 \cos \alpha + y_1 \cos \beta + z_1 \cos \gamma - p.}$$

That is, *to find the distance of the point P_1 from the plane $x \cos \alpha + y \cos \beta + z \cos \gamma = p$, substitute the coordinates of P_1 for x, y, and z in the expression $x \cos \alpha + y \cos \beta + z \cos \gamma - p$.*

270. COROLLARY. *The distance of the point $P_1(x_1, y_1, z_1)$ from the plane $Ax + By + Cz + D = 0$ is*

$$\mathbf{d = \frac{Ax_1 + By_1 + Cz_1 + D}{\pm\sqrt{A^2 + B^2 + C^2}},}$$

where the sign of the radical is the same as that of D.

When the point P_1 is on the side of the plane away from the origin O, then d has the same direction as p and is therefore positive. When P_1 is on the same side of the plane as the origin O, then d is negative. The sign may be neglected if simply the numerical distance is required.

Exercise 66. Equations of Planes

Find the equation of each plane determined as follows:

1. Perpendicular to a line having direction cosines proportional to 2, 3, -6, and 7 units from the origin.

2. Passing through the point $(4, 3, -6)$ and perpendicular to a line having direction cosines proportional to 1, 2, 1.

3. Passing through the point $(1, -2, 0)$ and perpendicular to the line determined by the points $(2, -4, 2)$ and $(-1, 3, 7)$.

4. Passing through the points $(1, 2, -2)$, $(0, 6, 2)$, $(3, 4, -5)$.

5. Passing through the point $(2, 2, 7)$ and having equal intercepts on the axes.

6. Passing through the points $(2, 3, 0)$ and $(1, -1, 2)$ and parallel to the z axis.

7. Parallel to the plane $4x - 2y - z = 12$ and passing through the origin.

8. Parallel to the plane $x - y + z = 7$ and passing through the point $(2, 3, 3)$.

9. Perpendicular to the line P_1P_2 at its mid point, where P_1 is the point $(4, 7, -3)$ and P_2 is the point $(-2, 1, -9)$.

10. Parallel to the plane $2x - 6y + 3z = 17$ and 9 units from the origin.

11. Parallel to the plane $2x - 6y - 3z = 21$ and 3 units further from the origin.

12. Tangent at $P(4, -12, 6)$ to the sphere with center at the origin O and with radius OP.

13. Passing through the point $(-2, -4, 3)$, 2 units from the origin O, and parallel to OY.

14. Passing through the point $(3, -2, 1)$, having the z intercept 7, and having the x intercept 7.

The intercept equation (§ 263) is a simple basis for the solution.

15. The equation of a plane cannot be written in the intercept form when the plane either passes through the origin or is parallel to an axis.

16. The equation of the xy plane is $z = 0$, that of the zx plane is $y = 0$, and that of the yz plane is $x = 0$.

17. What is the locus of all points (x, y, z) for which $x = 0$ and $y = 0$? for which $x = 0$ and $z = 0$? for which $y = 0$ and $z = 0$? Prove each statement.

18. Where are all points (x, y, z) for which $x = 3$ and $z = 2$? Prove the statement.

19. Find the distance of the point $(4, 3, 3)$ from the z axis.

20. Find the distance of the point $(3, -2, 3)$ from the plane $6x - 2y - 3z + 8 = 0$.

21. The vertex of a pyramid is the point $(2, 7, -2)$, and the base, lying in the plane $2x - 5y + z - 12 = 0$, has an area of 32 square units. Find the volume of the pyramid.

22. The vertices of a tetrahedron are $(1, 1, 1)$, $(3, -2, 1)$, $(-3, 4, 2)$, and $V(10, -8, 3)$. Find the length of the perpendicular from V upon the opposite face.

23. Find the value of k for which the plane $2x - y + z + k = 0$ passes at the distance 4 from the point $(3, -1, 7)$.

24. Find the distance from the plane $3x - 4y + 2z = 10$ to the parallel plane $3x - 4y + 2z = 30$.

25. Find the equation of the locus of points equidistant from the origin and the plane $6x + 2y - 3z = 8$.

26. The direction cosines of a line perpendicular to a plane are proportional to the reciprocals of the intercepts of the plane.

27. The sum of the reciprocals of the intercepts of planes through a fixed point whose coordinates are equal to each other is constant.

28. Find the volume of the tetrahedron formed by the three coordinate planes and the plane $Ax + By + Cz = D$.

THEOREM. PLANES THROUGH A LINE

271. *If the equations of two planes M and N which intersect in l are* $Ax + By + Cz + D = 0$ *and* $A'x + B'y + C'z + D' = 0$, *then* $Ax + By + Cz + D + k(A'x + B'y + C'z + D') = 0$, *k being an arbitrary constant, is the equation of the system of planes containing l.*

Proof. The last equation, being of the first degree, represents a plane R and is obviously satisfied by the coordinates of every point which is on both planes M and N. Hence the plane R contains l.

Furthermore, every plane which contains l is represented by the last equation. For if the point $P_1(x_1, y_1, z_1)$ is any point in space, the substitution of x_1, y_1, z_1 for x, y, z respectively leaves k as the only unknown quantity in the equation, and therefore k can be determined in such a way that the plane R passes through the point P_1.

In the special case when P_1 is on the first plane, we see that $k = 0$. But when P_1 is on the second plane, there is no finite value of k; and in this case it is sometimes said that $k = \infty$.

272. Straight Lines. Any straight line in space may be regarded as the intersection of the two planes

$$Ax + By + Cz + D = 0$$

and

$$A'x + B'y + C'z + D' = 0.$$

Therefore the coordinates of every point (x, y, z) on the line satisfy both of these equations; and, conversely, every point (x, y, z) whose coordinates satisfy both equations is a point on the line.

We therefore say that a straight line in space is represented by two equations of the first degree, regarded as simultaneous.

PROBLEM. SYMMETRIC EQUATIONS OF A LINE

273. *To find the equations of the line passing through the point $P_1(x_1, y_1, z_1)$ and having given direction angles.*

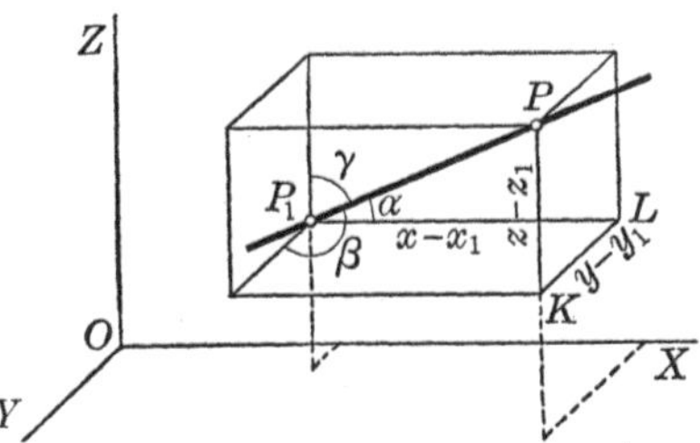

Solution. Let $P(x, y, z)$ be any point on the line. Draw the rectangular parallelepiped having PP_1 as diagonal and edges parallel to the axes, and let $PP_1 = r$.

Then $P_1L = x - x_1$, $LK = y - y_1$, $KP = z - z_1$, and (§ 251)

$$x - x_1 = r\cos\alpha, \quad y - y_1 = r\cos\beta, \quad z - z_1 = r\cos\gamma;$$

that is,

$$\frac{x - x_1}{\cos\alpha} = \frac{y - y_1}{\cos\beta} = \frac{z - z_1}{\cos\gamma}, \tag{1}$$

each fraction being equal to r.

These equations are called the *symmetric equations* of a line.

274. COROLLARY 1. *The equations of the line through $P_1(x_1, y_1, z_1)$ with direction cosines proportional to a, b, c are*

$$\frac{x - x_1}{a} = \frac{y - y_1}{b} = \frac{z - z_1}{c}. \tag{2}$$

This comes by multiplying the denominators of (1) by a constant.

275. COROLLARY 2. *The equations of the line passing through the points $P_1(x_1, y_1, z_1)$ and $P_2(x_2, y_2, z_2)$ are*

$$\frac{x - x_1}{x_2 - x_1} = \frac{y - y_1}{y_2 - y_1} = \frac{z - z_1}{z_2 - z_1}.$$

For $\cos\alpha : \cos\beta : \cos\gamma = x_2 - x_1 : y_2 - y_1 : z_2 - z_1$, by § 251.

276. Equations of a Line reduced to the Symmetric Form. The symmetric equations of a line (§ 273) may be regarded as expressing one of the variables x, y, z in terms of each of the others. This suggests a method for reducing the general equations of a line (§ 272) to the symmetric form, as in the illustration below.

For example, reduce to the symmetric form the following equations of a certain line:

$$7x - 2y + 3z = 2,$$
$$11x - 6y - 6z = -24, \qquad (1)$$

Eliminating z, we have $25x - 10y = -20$, whence $x = \frac{2}{5}(y-2)$. Eliminating y, we have $x = -\frac{3}{2}(z-2)$. These two new equations may be written $x = \frac{2}{5}(y-2) = -\frac{3}{2}(z-2)$, and they take the symmetric form when we divide each fraction by the number which removes the coefficients of the variables, thus:

$$x = \frac{y-2}{\frac{5}{2}} = \frac{z-2}{-\frac{2}{3}};$$

or, multiplying the denominators by 6,

$$\frac{x}{6} = \frac{y-2}{15} = \frac{z-2}{-4}. \qquad (2)$$

This is the symmetric form. The direction cosines of the line are proportional to 6, 15, -4.

The student should observe that each of the three equations in (2) involves only two of the variables x, y, z, and therefore represents a plane parallel to an axis (§ 268).

If a line is parallel to one of the coordinate planes, it is perpendicular to an axis; that is, if it is parallel to the xy plane it is perpendicular to OZ, and hence it follows that $\cos\gamma = 0$. In such a case the equation of the line cannot be written in symmetric form, for $\cos\gamma$ appears in a denominator in the symmetric equations (§ 273), and a fraction cannot have 0 as a denominator.

Problem. Line through a Point Perpendicular to a Plane

277. *To find the equations of the line which passes through the point* $P_1(x_1, y_1, z_1)$ *and is perpendicular to the plane* $Ax + By + Cz + D = 0$.

Solution. Since the direction cosines of the line must be proportional to A, B, C (§ 262), the required equations are

$$\frac{x - x_1}{A} = \frac{y - y_1}{B} = \frac{z - z_1}{C}. \qquad \text{§ 273}$$

Problem. Angle between Two Lines

278. *To find the angle* θ *between two lines* l *and* l' *in terms of their direction angles* α, β, γ *and* α', β', γ'.

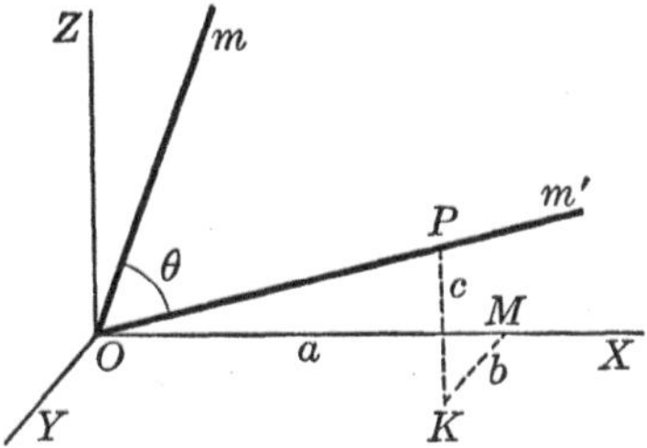

Solution. Draw through O the lines m and m' parallel to l and l' respectively. Draw the coordinates OM, MK, KP of any point $P(a, b, c)$ of m' and let $OP = r$.

Projecting the broken line $OMKP$ on m and noticing that the sum of the projections of OM, MK, and KP is equal to the projection of OP, we have

$$a \cos \alpha + b \cos \beta + c \cos \gamma = r \cos \theta, \qquad \text{§ 257}$$

that is,
$$r \cos \alpha' \cos \alpha + r \cos \beta' \cos \beta + r \cos \gamma' \cos \gamma = r \cos \theta, \qquad \text{§ 252}$$

or
$$\cos \theta = \cos \alpha \cos \alpha' + \cos \beta \cos \beta' + \cos \gamma \cos \gamma'.$$

279. Corollary 1. *The line l is perpendicular to the line l' if and only if*

$$\cos\alpha\cos\alpha' + \cos\beta\cos\beta' + \cos\gamma\cos\gamma' = 0.$$

For if l is perpendicular to l', $\cos\theta = 0$ (§ 278).

280. Corollary 2. *Let a, b, c be proportional to the direction cosines of l, and a', b', c' to those of l'. Then l is respectively perpendicular or parallel to l' if and only if*

$$aa' + bb' + cc' = 0 \quad \text{or} \quad a : a' = b : b' = c : c'.$$

For the direction cosines of l are

$$\frac{a}{\sqrt{a^2+b^2+c^2}}, \quad \frac{b}{\sqrt{a^2+b^2+c^2}}, \quad \frac{c}{\sqrt{a^2+b^2+c^2}}, \qquad \text{§ 253}$$

and those of l' are

$$\frac{a'}{\sqrt{a'^2+b'^2+c'^2}}, \quad \frac{b'}{\sqrt{a'^2+b'^2+c'^2}}, \quad \frac{c'}{\sqrt{a'^2+b'^2+c'^2}}.$$

We may now apply § 279 to prove the first part of this corollary.

In proving the second part the student should recall that l is parallel to l' if the direction cosines of l are either equal to those of l' or are the negatives of those of l', and not otherwise (§ 250).

281. Corollary 3. *The angle θ between the two planes $Ax + By + Cz + D = 0$ and $A'x + B'y + C'z + D' = 0$ is given by the equation*

$$\cos\theta = \pm\frac{AA' + BB' + CC'}{\sqrt{A^2+B^2+C^2}\cdot\sqrt{A'^2+B'^2+C'^2}}.$$

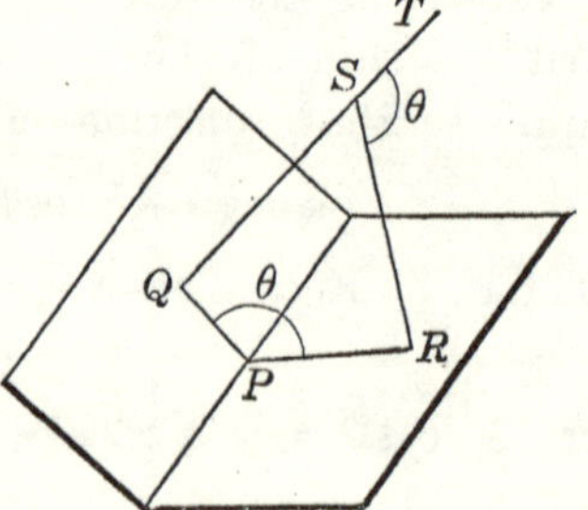

Through any point S draw QST perpendicular to one of the given planes, and draw RS perpendicular to the other. Let the plane QSR cut the intersection of the given planes at P. Then θ is equal to the angle RST, both being supplements of the angle QSR. Finding then the direction cosines of RS and ST by § 262, the formula follows by § 278.

Exercise 67. Straight Lines

Describe the line represented by each of the following pairs of equations:

1. $x = 4$, $y = 6$.
2. $y = -2$, $z = 1$.
3. $x = 0$, $y = 6$.
4. $y = 3$, $z = 0$.
5. $x = 0$, $y = 0$.
6. $x = 0$, $z = 0$.
7. $x = 1$, $z = 1$.
8. $y = 2$, $z = 0$.

Find the equations of the lines determined as follows:

9. Passing through the points $(4, 3, -2)$ and $(1, 1, 3)$.

10. Passing through the point $(2, 0, 1)$ and perpendicular to the plane $2x - y + 3z = 6$.

11. Passing through the point $(-1, 3, 3)$ and parallel to the line $\frac{1}{2}(x - 1) = \frac{1}{3}(y + 2) = \frac{1}{3}(z - 3)$.

12. Passing through the point $(2, 0, 0)$ and parallel to the line $2x + y + z = 10$, $x - 3y + z = 15$.

13. Find the direction cosines of the line $x + y = 5$, $y = z$.

14. The vertices of a tetrahedron are $A(2, 1, 0)$, $B(0, 4, 2)$, $C(-1, 2, -3)$, $D(6, -2, 2)$. Find the face angle ACB, and the angle between the planes which intersect along AC.

15. Find the angle formed by the two lines $2x + y - z = 7$, $x + y + z = 10$ and $x - 3y + 2z = 5$, $3x - 3y + z = 7$.

16. Find the angle formed by the two lines $x = y = z$ and $2x = y = 1 - z$.

17. The points $(1, -2, 0)$, $(10, -6, 6)$, $(-8, 2, -6)$ lie on one straight line.

18. Find the equations of the lines in which the coordinate planes are cut by the plane $2x + 5y - 7z = 8$.

19. Show how to find the coordinates of several points on a given line; for example, on the line $3x - y + z = 4$, $2x + 2y - 3z = 8$.

282. Other Systems of Coordinates. There are many systems of coordinates for locating points in space, other than the rectangular system which we have been employing, and two of the most important of these systems are here explained.

1. *Spheric Coordinates.* Let OX, OY, OZ be three lines each perpendicular to the other two, as in the rectangular system. Then any point P in space is located by the distance OP, or ρ, the angle θ which OP makes with the z axis, and the angle ϕ which the zx plane makes with the plane determined by OZ and OP. The three quantities ρ, θ, ϕ are called the *spheric coordinates* of the point P, and are denoted by (ρ, θ, ϕ).

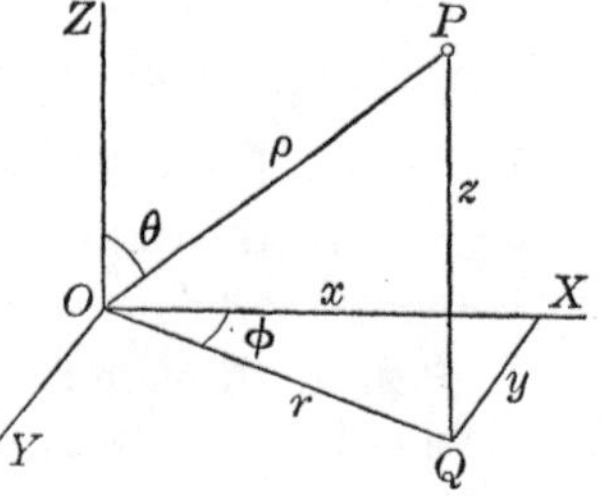

After examining carefully the above figure, the student will have no difficulty in showing that the rectangular coordinates x, y, z of the point P bear the following relations to the spheric coordinates ρ, θ, ϕ of P:

$$x = \rho \sin \theta \cos \phi, \quad y = \rho \sin \theta \sin \phi, \quad z = \rho \cos \theta.$$

For example, $x = OQ \cos \phi = \rho \cos (90° - \theta) \cos \phi$.

Spheric coordinates are employed for many investigations in astronomy, physics, and mechanics.

2. *Cylindric Coordinates.* Again, the point P may be located by the distance OQ, or r, the angle ϕ, and the distance z. The quantities r, ϕ, z are called the *cylindric coordinates* of P and are denoted by (r, ϕ, z).

Obviously, the cylindric coordinates are related to the rectangular coordinates x, y, z as follows:

$$x = r \cos \phi, \quad y = r \sin \phi, \quad z = z.$$

Exercise 68. Review

Draw the tetrahedron having the vertices $A(4, 8, 0)$, $B(2, 5, -2)$, $C(3, 2, 2)$, $D(5, 1, 2)$, and find the following:

1. The length and the direction cosines of AB.

2. The equations of AB and BC.

3. The cosine of the angle ABC.

4. The equation of the plane containing A, B, and C.

5. The altitude measured from D to the plane ABC.

6. The equations of the line through D perpendicular to the plane ABC.

7. The cosine of the dihedral angle having the edge AB.

8. The equations of the line perpendicular to the face ABC at the point of intersection of the medians of that face.

9. The equation of the plane through A parallel to the plane BCD.

10. The equation of the plane through C perpendicular to AB.

11. Find the equations of the line through the origin perpendicular to the plane $3x - 2y + 7z = 10$.

12. Find the equations of the line through the point (2, 1, 3) parallel to the line $2x - y + 2z = 5$, $x - 3y + 3z = 8$.

13. Find the coordinates of the point in which the line through the points $(2, 2, -4)$ and $(6, 0, 2)$ cuts the plane $6x - 8y + 2z = 4$.

14. Find the coordinates of each of the points in which the line $3x + y - 3z = 10$, $2x + 2y - z = 4$ cuts the three coordinate planes.

15. Find the coordinates of the points in which the line $x - 4 = \frac{y+2}{-4} = \frac{z-9}{3}$ cuts the sphere $x^2 + y^2 + z^2 = 49$.

16. Find the common point of the planes $x + y + z = 4$, $x = y + z$, and $3x - y + z = 6$.

17. The planes $2x + y + z - 10 = 0$, $x - y + 3z - 14 = 0$, $4x - y + 7z - 20 = 0$ have no common point but meet by pairs in parallel lines.

18. Find the equation of the plane that contains the line $x + y = z$, $3x + 2y + 2 = 10$ and the point $(6, -2, 2)$.

19. Find the equation of the plane that contains the line $x = y = z - 6$ and is $\sqrt{6}$ units from the origin.

20. Find the equations of the planes which bisect the angles between the planes $2x - y + z = 5$ and $x + y - 2z = 8$.

21. The direction cosines of the line determined by the planes $Ax + By + Cz + D = 0$, $A'x + B'y + C'z + D' = 0$ are proportional to $BC' - B'C$, $AC' - A'C$, $AB' - A'B$.

22. The planes $Ax + By + Cz + D = 0$, $A'x + B'y + C'z + D = 0$ are perpendicular to each other if $AA' + BB' + CC' = 0$, and conversely.

23. Show that the line $\frac{1}{2}(x - 2) = \frac{1}{3}(y + 1) = \frac{1}{3}z$ intersects the line $\frac{1}{4}(x - 2) = \frac{1}{5}(y + 1) = \frac{1}{2}z$, and find the equation of the plane determined by them.

24. The equation of the plane through the points (x_1, y_1, z_1), (x_2, y_2, z_2), (x_3, y_3, z_3) may be written in the determinant form here shown.

$$\begin{vmatrix} x & y & z & 1 \\ x_1 & y_1 & z_1 & 1 \\ x_2 & y_2 & z_2 & 1 \\ x_3 & y_3 & z_3 & 1 \end{vmatrix} = 0.$$

Ex. 24 should be omitted by those who have not studied determinants.

25. In spheric coordinates the equation of the sphere with center O and radius r is $\rho^2(\sin^2\phi + \cos^2\theta) = r^2$.

Find the equation of the locus of the point P which satisfies the conditions in each of the following cases:

26. The ratio of the distances of P from $A(2, 0, 0)$ and $B(-4, 0, 0)$ is a constant.

27. The distance of P from the zx plane is equal to the distance from P to $A(4, -2, 1)$.

CHAPTER XIV

SURFACES

283. Sphere. If $P(x, y, z)$ is any point of the surface of the sphere having the center $C(a, b, c)$ and the radius r, then $CP = r$; that is, by § 248,

$$(x - a)^2 + (y - b)^2 + (z - c)^2 = r^2. \qquad (1)$$

This equation may also be written

$$x^2 + y^2 + z^2 - 2\,ax - 2\,by - 2\,cz + d = 0, \qquad (2)$$

where $d = a^2 + b^2 + c^2 - r^2$. It is an equation of the second degree in which the coefficients of x^2, y^2, and z^2 are equal, and it contains no xy term, yz term, or xz term.

Every equation of the form (2) can be reduced to the form (1) by a process of completing squares, and it represents a sphere having the center (a, b, c) and the radius r.

For example, the equation $x^2 + y^2 + z^2 + 6\,x - 4\,y - 4\,z = 8$ can be written $(x + 3)^2 + (y - 2)^2 + (z - 2)^2 = 25$, and represents a sphere with center $(-3, 2, 2)$ and radius 5.

As in the case of the circle (§ 91), we have a *point sphere* if $r = 0$, or an *imaginary sphere* if r^2 is negative.

If the center of the sphere is the origin (0, 0, 0), the equation of the sphere is evidently

$$x^2 + y^2 + z^2 = r^2.$$

We shall hereafter use the term *sphere* to mean the surface of the sphere; and similarly in the cases of the cylinder, cone, and other surfaces studied.

284. Cylinder Parallel to an Axis. In plane analytic geometry the equation $x^2 + y^2 = r^2$ represents a circle. We shall now investigate the locus in space of a point whose coordinates satisfy the equation $x^2 + y^2 = r^2$.

In the xy plane draw the circle with center O and radius r. The equation of this circle in its plane is $x^2 + y^2 = r^2$. Let $P'(x, y, 0)$ be any point on the circle. Through P' draw a line parallel to the z axis, and let $P(x, y, z)$ be any point on this line. Then the coordinates of P' obviously satisfy the equation $x^2 + y^2 = r^2$.

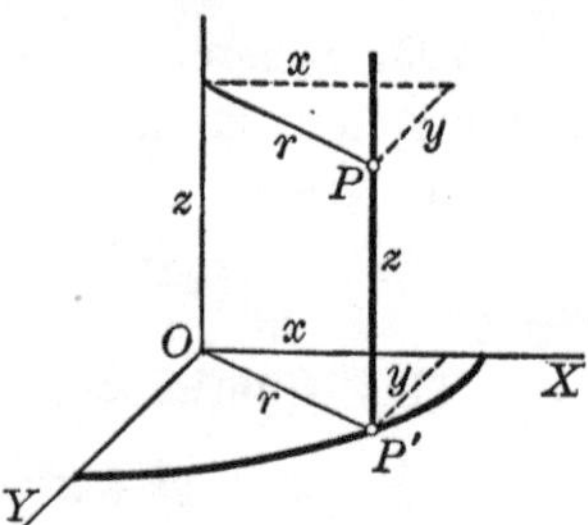

But the x and y of P are the same as those of P', and hence the equation $x^2 + y^2 = r^2$ is true for P and is also true for every point of the line $P'P$. Further, since P' is any point on the circle, the equation $x^2 + y^2 = r^2$ is true for every point of the circular cylinder whose elements are parallel to the z axis and pass through the circle, and for no other points.

Similarly, the equation $2x^2 + 3y^2 = 18$ represents the cylinder whose elements are parallel to the z axis and pass through the ellipse $2x^2 + 3y^2 = 18$ in the xy plane.

In general, then, we see that

Any equation involving only two of the variables x, y, z represents a cylinder parallel to the axis of the missing variable and passing through the curve determined by the given equation in the corresponding coordinate plane.

As a further example, the equation $y^2 = 6z$ represents the cylinder parallel to the x axis and passing through the parabola $y^2 = 6z$ in the yz plane.

The term *cylinder* is used to denote any surface generated by a straight line moving parallel to a fixed straight line.

285. Trace of a Surface. The curve in which a surface cuts a coordinate plane is called the *trace* of the surface in that plane.

Since $z = 0$ for all points in the xy plane and for no others, the equation of the xy trace of a surface is found by letting $z = 0$ in the equation of the surface.

By the xy trace of a surface is meant the trace of the surface in the xy plane. Thus the xy trace of the plane $3x - y - z = 1$ is the line $3x - y = 1$, $z = 0$.

The equation of the zx trace of a surface is found by letting $y = 0$ in the equation of the surface, and the equation of the yz trace is found by letting $x = 0$.

286. Contour of a Surface. The curve in which a plane parallel to the xy plane cuts a surface is called an *xy contour* of the surface; and, similarly, we have *yz contours* and *zx contours*.

To find the xy contour which the plane $z = k$ cuts from a surface, we evidently must let $z = k$ in the equation of the surface. To find the zx contour made by the plane $y = k$, we let $y = k$ in the equation; and for the yz contour made by the plane $x = k$, we let $x = k$ in the equation.

For example, the xy contour of the sphere $x^2 + y^2 + z^2 = 25$ made by the plane $z = 2$ is the circle $x^2 + y^2 = 21$, of which DE is one quadrant.

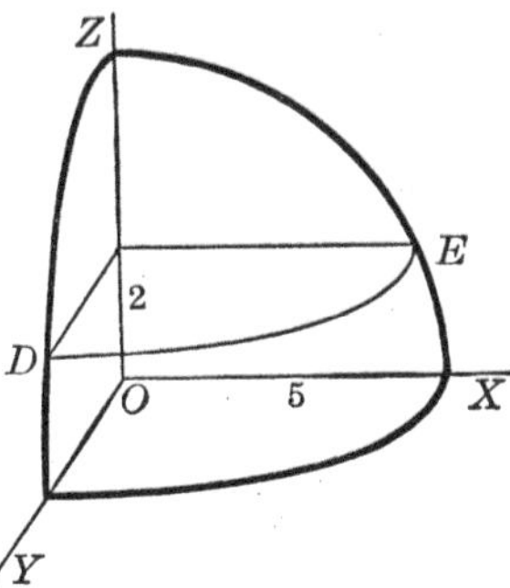

Observe that this circle is not represented by one equation, but by the two equations $x^2 + y^2 = 21$, $z = 2$, or we may speak of the circle $x^2 + y^2 = 21$ in the plane $z = 2$. The single equation $x^2 + y^2 = 21$ represents in space a cylinder parallel to OZ (§ 284).

It is not necessary to draw more than one octant of the sphere.

287. Cone having the Origin as Vertex. The equation

$$\frac{x^2}{a^2}+\frac{y^2}{b^2}-\frac{z^2}{c^2}=0 \qquad (1)$$

is satisfied by the coordinates 0, 0, 0 of the origin O. Let x', y', z' be any other numbers which satisfy equation (1). Then it is obvious that the numbers kx', ky', kz', where k is any constant whatever, must also satisfy equation (1).

But $P(kx', ky', kz')$ represents any and every point of the straight line determined by the origin and the point $P'(x', y', z')$.

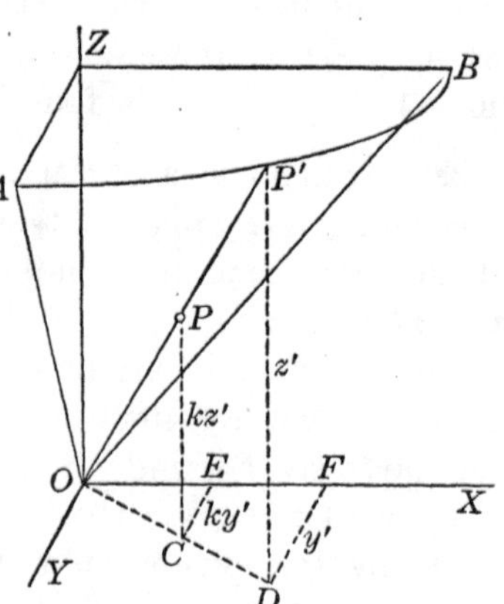

This may be seen directly from a figure, or by using the results of § 254 and taking P_1 and P_2 of § 254 as (0, 0, 0) and (x', y', z') in the present discussion.

Hence all points of the line determined by the origin and any point of the locus of (1) lie on the locus.

Moreover, the section of the locus made by the plane $z=c$ is the ellipse

$$\frac{x^2}{a^2}+\frac{y^2}{b^2}=1,\ z=c. \qquad (2)$$

Therefore the locus of (1) is the cone having the origin as vertex and passing through the ellipse (2).

The above method of reasoning applies also to the equations $\frac{x^2}{a^2}\pm\frac{y^2}{b^2}\pm\frac{z^2}{c^2}=0$, to $y^2=axz$, and in fact to any homogeneous equation in x, y, z; that is, to any equation in which all terms are of the same degree. Every such equation represents a cone having the origin as vertex.

Exercise 69. Spheres, Cylinders, Cones

Find the equations of the spheres determined as follows:

1. Passing through the point $(4, -2, 2)$ and having its center at the point $(2, 1, -4)$.

2. Center $(4, -4, 3)$, and tangent to the xy plane.

3. Tangent to the plane $6x - 2y - 3z = 28$ and having its center at the point $(6, 5, -1)$.

4. Tangent to the coordinate planes and passing through the point $(5, 10, 1)$.

5. Passing through the four points $(3, 2, -4)$, $(-2, -3, 8)$, $(-5, 2, 4)$, $(4, -7, 0)$.

6. Having as diameter the line joining $P_1(4, -1, 3)$ and $P_2(-8, 3, -3)$.

7. Radius 10, center on the line $x + y + z = 4$, $x = y + z$, and tangent to the plane $3x - 2y + 6z = 21$.

Draw the following cylinders and cones:

8. $x^2 + y^2 = 25$.

9. $x^2 + z^2 = 25$.

10. $y^2 = 12x$.

11. $x^2 - y^2 = 16$.

12. $x^2 + y^2 = 6x$.

13. $x^2 + z^2 - 2x - 4z = 4$.

14. $4x^2 - 25y^2 - 100z^2 = 0$.

15. $y^2 = 4xz$.

Find the traces of the following surfaces on the coordinate planes, and find the equations of the sections made by the planes parallel to the coordinate planes and 1 unit from them, drawing each trace and section:

16. $x^2 + y^2 + z^2 = 20x$.

17. $z^2 = 9xy$.

18. $y^2 + 4z^2 = 4$.

19. $z = x^2 + y^2$.

20. $z = xy$.

21. $z = 4x^2 + 9y^2$.

22. Find the center and also the radius of the sphere $x^2 + y^2 + z^2 + 6x - 8y + 2z = 10$.

288. Quadric Surface or Conicoid. Any surface which is represented by an equation of the second degree in x, y, and z is called a *quadric surface* or a *conicoid*.

The sphere, cylinder, and cone are familiar conicoids. We have already discussed these surfaces and shall now discuss other conicoids.

289. Ellipsoid. Let us first consider the equation

$$\frac{x^2}{a^2}+\frac{y^2}{b^2}+\frac{z^2}{c^2}=1.$$

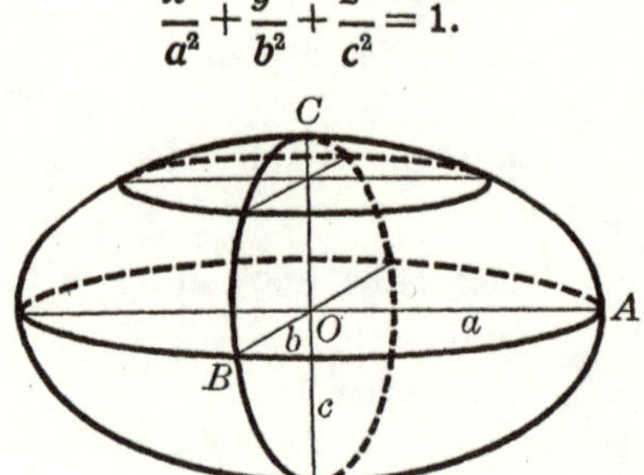

The xy trace, yz trace, and zx trace of the locus of the given equation are all ellipses. § 285

The student should write the equation of each trace and should draw the trace. Thus the xy trace is $x^2/a^2 + y^2/b^2 = 1$.

The xy contour made by the plane $z = k$ is the ellipse

$$\frac{x^2}{a^2}+\frac{y^2}{b^2}=1-\frac{k^2}{c^2}.$$

This contour is largest when $k = 0$; it decreases as k increases numerically; it becomes a point when $k = c$; and it is imaginary when $k > c$.

The yz contours are ellipses which vary in a similar manner, and so are the zx contours.

The intercepts of the locus on the axes are $\pm a$ on OX, $\pm b$ on OY, and $\pm c$ on OZ.

The surface is called an *ellipsoid*.

290. Simple Hyperboloid. Let us consider the equation

$$\frac{x^2}{a^2}+\frac{y^2}{b^2}-\frac{z^2}{c^2}=1.$$

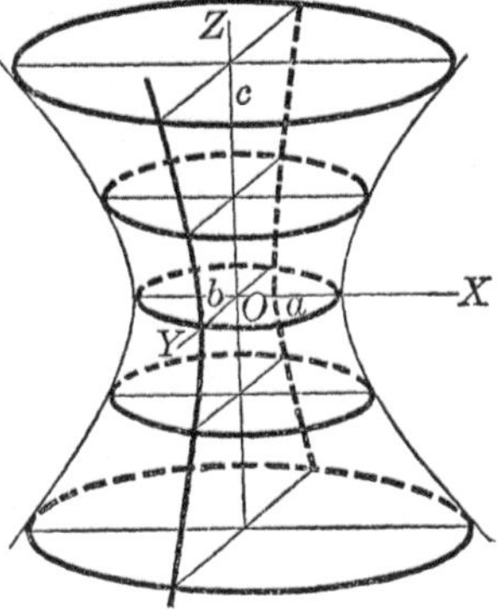

The xy trace of the locus of this equation is the ellipse

$$\frac{x^2}{a^2}+\frac{y^2}{b^2}=1.$$

The yz trace and zx trace are hyperbolas.

The xy contour made by the plane $z=k$ is an ellipse, the smallest ellipse being formed when $k=0$; the ellipse increases without limit as k increases without limit.

The above facts are sufficient to make clear the shape of this surface, which is called a *simple hyperboloid*, or a *hyperboloid of one sheet.*

The student should discuss the yz contours and the zx contours.

291. Asymptotic Cone. The cone $\frac{x^2}{a^2}+\frac{y^2}{b^2}-\frac{z^2}{c^2}=0$ is called the *asymptotic cone* of the hyperboloid.

That the hyperboloid approaches this cone as an asymptote may be seen by examining the traces of the cone and the hyperboloid on the yz plane and on the zx plane.

292. Double Hyperboloid. Let us consider the equation

$$\frac{x^2}{a^2} - \frac{y^2}{b^2} - \frac{z^2}{c^2} = 1.$$

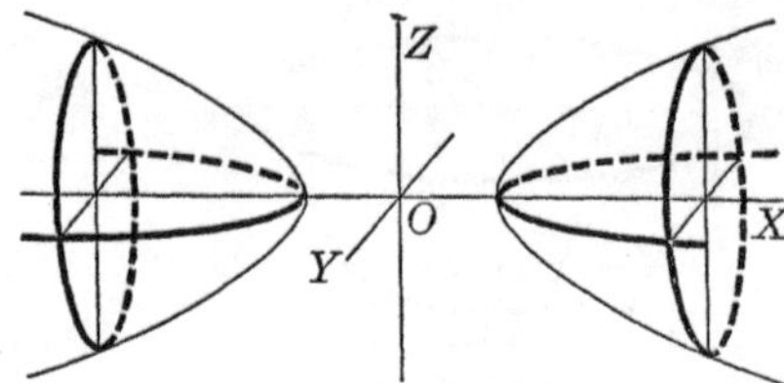

The xy trace of the locus of this equation is the hyperbola

$$\frac{x^2}{a^2} - \frac{y^2}{b^2} = 1.$$

The zx trace is the hyperbola

$$\frac{x^2}{a^2} - \frac{z^2}{c^2} = 1.$$

The yz trace is imaginary, being the imaginary hyperbola

$$\frac{y^2}{b^2} + \frac{z^2}{c^2} = -1.$$

The yz contour made by the plane $z = k$ is the ellipse

$$\frac{y^2}{b^2} + \frac{z^2}{c^2} = \frac{k^2}{a^2} - 1.$$

This ellipse is imaginary when $|k| < a$; it becomes a single point when $|k| = a$; and it is real when $|k| > a$, both major and minor axes increasing without limit as $|k|$ does.

The shape of the surface is now evident. The surface is called a *double hyperboloid*, or a *hyperboloid of two sheets.*

The student may discuss the xy contours and the zx contours of the surface and show that the xy trace and the zx trace are asymptotic to the corresponding traces of the cone $\frac{x^2}{a^2} - \frac{y^2}{b^2} - \frac{z^2}{c^2} = 0.$

293. Elliptic Paraboloid. Let us consider the equation

$$\frac{x^2}{a^2} + \frac{y^2}{b^2} = 2\,cz.$$

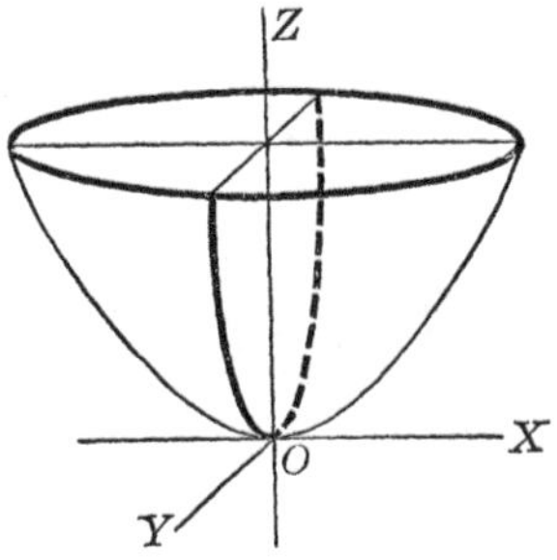

The equation of the xy trace of the locus is evidently

$$\frac{x^2}{a^2} + \frac{y^2}{b^2} = 0,$$

which represents a single point (0, 0) in the xy plane.

The yz trace is evidently the parabola

$$y^2 = 2\,b^2cz.$$

The zx trace is evidently the parabola

$$x^2 = 2\,a^2cz.$$

The xy contour made by the plane $z = k$ is the ellipse

$$\frac{x^2}{a^2} + \frac{y^2}{b^2} = 2\,ck,$$

its axes increasing without limit when k does. The ellipse is imaginary when k is negative.

The above statements indicate the shape of the surface, which is called an *elliptic paraboloid.*

If $c < 0$, the surface is below the xy plane.

The student may show that the zx contour is a parabola which remains constant; and similarly for the yz contour.

294. Hyperbolic Paraboloid. Consider the equation

$$\frac{x^2}{a^2}-\frac{y^2}{b^2}=2\,cz.$$

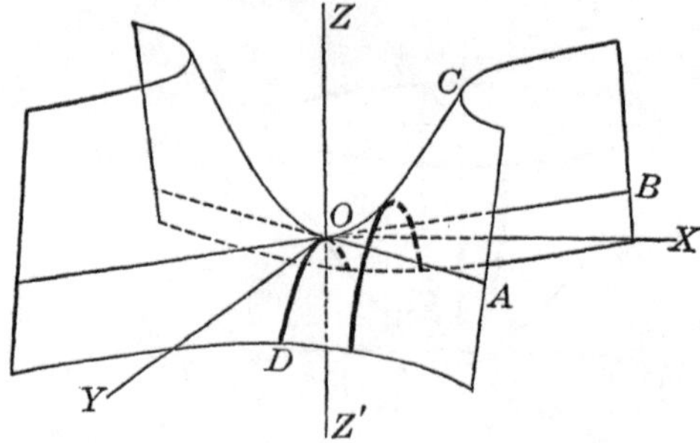

The equation of the xy trace of the locus is evidently

$$\frac{x^2}{a^2}-\frac{y^2}{b^2}=0,$$

which represents the two straight lines OA and OB (§ 54).

The yz trace is the parabola $y^2=-2\,b^2cz$, or OD.

The zx trace is the parabola $x^2=2\,a^2cz$, or OC.

The yz contour made by the plane $x=k$ is the parabola $y^2=-2\,b^2cz+b^2k^2/a^2$. This parabola remains constant in form, since it always has the focal width $2\,b^2c$. Also, if $c>0$, the axis of the parabola has the negative direction of the z axis.

The locus is symmetric with respect to the zx plane; for if x, y, z satisfy the given equation, so do x, $-y$, z.

The surface may therefore be regarded as generated by a parabola of constant size moving with its plane parallel to the yz plane, its axis parallel to OZ', and its vertex on the fixed parabola $x^2=2\,a^2cz$ in the zx plane.

The surface is called a *hyperbolic paraboloid*, the xy contours being hyperbolas.

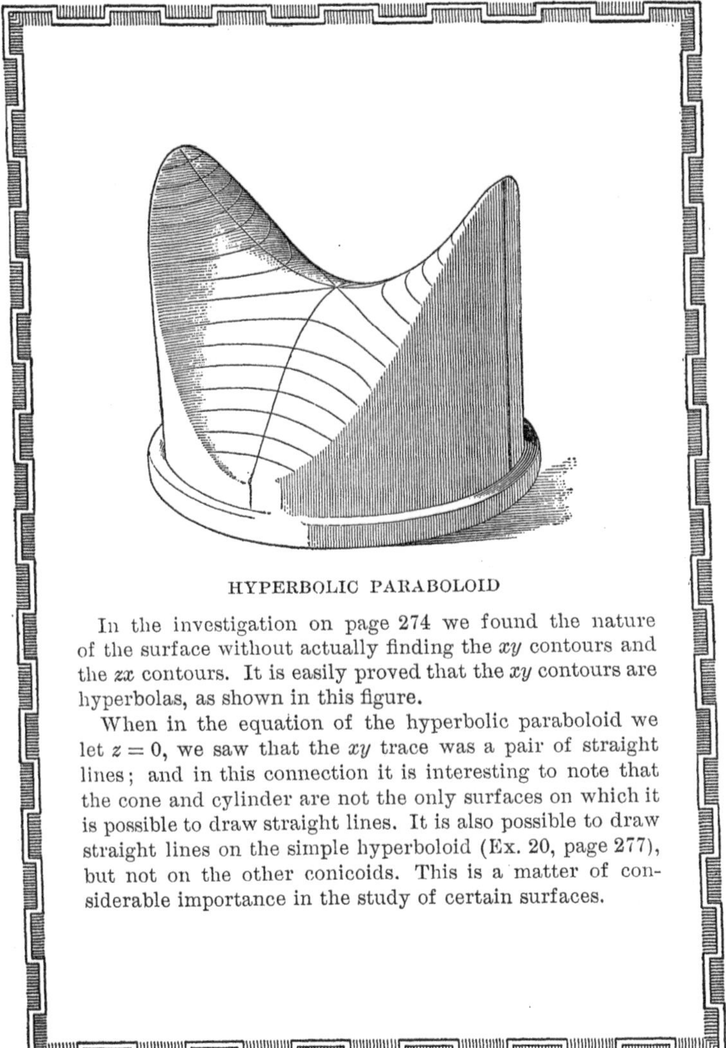

HYPERBOLIC PARABOLOID

In the investigation on page 274 we found the nature of the surface without actually finding the xy contours and the zx contours. It is easily proved that the xy contours are hyperbolas, as shown in this figure.

When in the equation of the hyperbolic paraboloid we let $z = 0$, we saw that the xy trace was a pair of straight lines; and in this connection it is interesting to note that the cone and cylinder are not the only surfaces on which it is possible to draw straight lines. It is also possible to draw straight lines on the simple hyperboloid (Ex. 20, page 277), but not on the other conicoids. This is a matter of considerable importance in the study of certain surfaces.

295. General Equation of the Second Degree. The general equation of the second degree in x, y, and z is

$$ax^2 + by^2 + cz^2 + dxy + eyz + fzx + gx + hy + iz + j = 0.$$

By transformations of coordinates analogous to those which we employed in Chapters VI and X, this equation can always be reduced to one of the two forms

$$a'x^2 + b'y^2 + c'z^2 = d', \qquad (1)$$

$$a'x^2 + b'y^2 = c'z. \qquad (2)$$

The theory of transformations in three dimensions, and the reductions referred to in the above statement, will not be given in this book.

The ellipsoid and hyperboloids are included under (1), and the paraboloids under (2). These are the only surfaces which are represented by equations of the second degree, and are called conicoids (§ 288). They include as special cases the sphere, cylinder, and cone.

296. Locus of any Equation. The nature of the locus of a point whose coordinates satisfy any given equation in x, y, and z may be investigated by the method which we have employed in §§ 289–294; that is, by examining the traces and contours of the locus.

It is evident that these traces and contours are curves of some kind, since when any constant k is substituted for x in the equation, we obtain an equation in two variables, and this equation represents a curve in the plane $x = k$.

We then see that

Every equation in rectangular coordinates represents a surface in space.

It is often difficult to form in the mind a clear picture of the surface represented by an equation, even after the traces and contours have been studied.

Exercise 70. Loci of Equations

1. Sketch the xy trace, the yz trace, and the zx trace of the surface $z = x^2 + y^2 - 9$. Discuss the xy contour, and represent the surface by a figure.

2. Sketch the xy trace, the yz trace, and the zx trace of the surface $x^2 - 4y^2 + z^2 = 36$. Choose one or more sets of contours, discuss each, and represent the surface by a figure.

As in Ex. 2, sketch and discuss the following surfaces:

3. $x^2 + y^2 + 16z^2 = 16$.

4. $x^2 + y^2 + 16z^2 = 25$.

5. $9x^2 + 4z^2 = 36y$.

6. $y^2 - 4z^2 = x^2 - 4$.

7. $4x - y^2 - z^2 = -4$.

8. $x^2 + y^2 + 25 = z^2$.

9. $y^2 = 12x - 36$.

10. $z^2 = 2xy$.

11. $xy = 10z$.

12. $y^2 = 4x - 4z^2$.

13. $x^2 + y^2 = z^2$.

14. $y^2 + z^2 = 8z$.

15. $z^2 = x - y$.

16. $yz = 10$.

17. Find the equation of the locus of a point which is equidistant from the point (8, 0, 0) and the plane $x = -8$, and sketch the surface.

18. Find the equation of the locus of a point P which moves so that the sum of the distances of P from the points (1, 0, 0) and (– 1, 0, 0) is 6, and sketch the surface.

19. Find the equation of the locus of a point P which moves so that the distances of P from the plane $x = y$ and the point (3, 1, 1) are equal.

20. Show that the plane $y = b$ cuts the simple hyperboloid in two straight lines.

The student should let the equation of § 290 represent the surface.

21. Find the equation of an ellipsoid having the contours $x^2 + 4y^2 = 16$, $z = 1$ and $16y^2 + z^2 = 64$, $x = 1$.

Let the equation $px^2 + qy^2 + rz^2 = 1$ represent the ellipsoid.

PROBLEM. SURFACE OF REVOLUTION

297. *To find the equation of the surface generated by a given curve revolving about the x axis.*

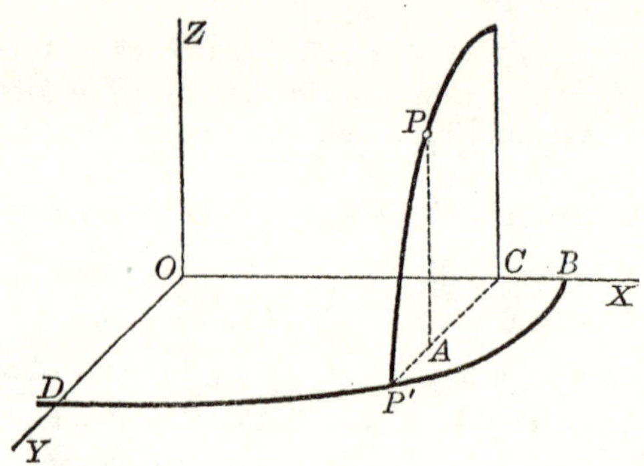

Solution. Let BD be the given curve in the xy plane. Let its equation, when solved for y in terms of x, be represented by $y=f(x)$, and let P' be any point of the curve. Since CP' is the y of P', then $CP'=f(x)$. § 62

As the curve revolves about OX, the point P' generates a circle with center C and radius CP'. Let $P(x, y, z)$ be any point on this circle. Then P is any point on the surface, and obviously

$$y^2+z^2=\overline{CP}^2=\overline{CP'}^2;$$

that is,

$$\mathbf{y^2+z^2=[f(x)]^2.}$$

For example, if the given curve is $2x^2+y^2=2$, then we find $y=\sqrt{2-2x^2}=f(x)$. This ellipse revolving about OX generates the surface $y^2+z^2=(\sqrt{2-2x^2})^2$, or $2x^2+y^2+z^2=2$.

298. COROLLARY. *The equations of the surfaces generated by revolving the curve $x=f(z)$, $y=0$ about the z axis and the curve $x=f(y)$, $z=0$ about the y axis are respectively*

$$\mathbf{x^2+y^2=[f(z)]^2} \quad \textit{and} \quad \mathbf{x^2+z^2=[f(y)]^2}.$$

Exercise 71. Surfaces of Revolution

1. Find the equation of the surface generated by revolving the ellipse $4x^2 + 9y^2 = 36$, $z = 0$ about the y axis.

In this case it is necessary to find x in terms of y. Solving for x, we have $x = \frac{1}{2}\sqrt{36 - 9y^2}$. Thus the required equation is $x^2 + z^2 = \frac{1}{4}(36 - 9y^2)$, or $4x^2 + 9y^2 + 4z^2 = 36$.

In each of the following examples find the equation of the surface which is generated by the given curve revolving about the axis specified:

2. The parabola $y^2 = 4px$, $z = 0$ about OX.

3. The parabola $y^2 = -4px$, $z = 0$ about OY.

This surface, having an equation of the fourth degree, is not a conicoid. The student may show that some of its contours are not conics.

4. The parabola $z^2 = 4px$, $y = 0$ about OX.

5. The parabola $z^2 = 4px$, $y = 0$ about OZ.

6. The ellipse $b^2x^2 + a^2y^2 = a^2b^2$, $z = 0$ about OY.

7. The ellipse $b^2x^2 + a^2z^2 = a^2b^2$, $y = 0$ about OZ.

8. The hyperbola $b^2x^2 - a^2y^2 = a^2b^2$, $z = 0$ about OX.

9. The hyperbola $b^2x^2 - a^2y^2 = a^2b^2$, $z = 0$ about OY.

10. The circle $x^2 + y^2 = 25$, $z = 0$ about OX; about OY.

11. The line $y = mx$, $z = 0$ about OY.

12. The circle $x^2 + y^2 - 12x + 32 = 0$, $z = 0$ about OX.

13. The circle $x^2 + y^2 - 12x + 32 = 0$, $z = 0$ about OY.

This surface is not a conicoid. It has the shape of a bicycle tire and is called a *torus* or *anchor ring*.

14. The curve $y = \sin x$, $z = 0$ about OX.

15. The curve $z = e^x$, $y = 0$ about OZ.

16. A parabolic headlight reflector is 8 in. across the face and 8 in. deep. Find the equation of the reflecting surface.

17. Find the equation of the cone generated by revolving the line $y = mx + b$, $z = 0$ about OX.

299. Equations of a Curve. The points common to two surfaces form a curve in space. The coordinates of every point on this curve satisfy the equation of one surface and also satisfy the equation of the other surface.

A curve in space, therefore, has two equations, regarded as simultaneous; namely, the equations of two surfaces of which the curve is the intersection.

For example, the equations $z = 2x + y$ and $x^2 + y^2 + z^2 = 25$ represent separately a plane and a sphere. Regarded as simultaneous, they are the equations of the circle formed by the intersection of the plane and the sphere.

300. Elimination of One Variable. If we eliminate one variable, z for example, from the two equations of a curve, we obtain a third equation which contains the other two variables x and y. This equation represents a third surface, a cylinder (§ 284), which contains the curve. The xy trace of this cylinder is the projection of the curve upon the xy plane, and the cylinder is called the xy projecting cylinder of the curve.

For example, eliminating z from the equations $z = 2x + y$ and $x^2 + y^2 + z^2 = 25$ of the circle described in § 299, we obtain the equation $5x^2 + 2y^2 + 4xy = 25$. This equation represents a cylinder parallel to OZ. The trace of this cylinder on the xy plane is the curve $5x^2 + 2y^2 + 4xy = 25$, $z = 0$, which is an ellipse.

Eliminating x, we obtain the xy projecting cylinder; eliminating y, we obtain the zx projecting cylinder.

301. Parametric Equations of a Curve. The three equations $x = f(v)$, $y = g(v)$, $z = h(v)$, giving x, y, and z in terms of a fourth variable v, represent a curve in space. For eliminating v from the first two equations gives one equation in x and y; and eliminating v from the last two equations gives one equation in y and z. These two new equations define a curve in space (§ 299).

Exercise 72. Review

1. Find the equation of the plane tangent to the sphere $x^2 + y^2 + z^2 = r^2$ at the point (x', y', z'), and show that it can be reduced to the form $x'x + y'y + z'z = r^2$.

2. If the two spheres $S = 0$ and $S' = 0$ are given, where S denotes $x^2 + y^2 + z^2 - 2ax - 2by - 2cz + d$ and S' denotes $x^2 + y^2 + z^2 - 2a'x - 2b'y - 2c'z + d'$, show that the equation $S - S' = 0$ represents the plane which contains the circle of intersection of the spheres, and that this plane is perpendicular to the line of centers.

3. The square of the length of a tangent from the point (x', y', z') to the sphere $(x - a)^2 + (y - b)^2 + (z - c)^2 = r^2$ is $(x' - a)^2 + (y' - b)^2 + (z' - c)^2 - r^2$.

4. The hyperbola cut from the hyperbolic paraboloid $b^2x^2 - a^2y^2 = a^2b^2cz$ by the plane $z = k$ has its real axis parallel to OX when k is positive, but parallel to OY when k is negative.

5. The asymptotes of any xy contour of the hyperbolic paraboloid in Ex. 4 are parallel to the asymptotes of any other xy contour of the surface.

6. Show by means of traces and contours that the equation $(x - a)^2 + y^2 - z^2 = 0$ represents a right circular cone having the point $(a, 0, 0)$ as vertex.

7. Draw the yz trace of the surface $x^3 + y^3 + z^3 = 0$.

8. Find the equation of the cone which passes through the origin and which has its elements tangent to the sphere $x^2 + y^2 + z^2 - 20z + 36 = 0$.

9. Find the equation of the paraboloid having the origin as vertex and passing through the circle $x^2 + y^2 = 25$, $z = 3$.

10. A tank in the form of an elliptic paraboloid is 12 ft. deep and the top is an ellipse $8\sqrt{3}$ ft. long and 12 ft. wide. If the upper part of the tank is cut off parallel to the top 3 ft. below the top, find the axes of the ellipse forming the new top.

Given the points $A(4, 0, 0)$ and $B(-4, 0, 0)$, find the equations of the loci in Exs. 11–15, and state in each case what kind of surface the locus is:

11. The locus of a point P such that the ratio of the distances AP and BP is constant.

12. The locus of a point P such that the sum of the squares of the distances AP and BP is constant.

13. The locus of a point P such that the difference of the distances AP and BP is constant.

14. The locus of a point P such that the distance AP is equal to the distance of P from the plane $x = -4$.

15. The locus of a point P such that the distance AP is four fifths the distance of P from the plane $x = \frac{25}{4}$.

16. Draw the ellipsoid $4x^2 + 4y^2 + 25z^2 = 100$, x, y, and z being measured in inches, and draw a rectangular box having its vertices on the ellipsoid and its edges parallel to the axes. If the volume of the box is 72 cu. in. and the dimension parallel to the z axis is 2 in., find the other dimensions.

17. Find the length of the segment of the line $x = 3$, $y = 5$ intercepted between the plane $x - y + 3z = 10$ and the hyperboloid $3x^2 - 2y^2 - z^2 = -123$.

18. Find the coordinates of the points in which the line $x = y = z$ cuts the paraboloid $3x^2 + 4y^2 = 12z$.

19. Find the coordinates of the points in which the line through the points $(2, 0, 2)$ and $(3, 2, -1)$ cuts the sphere having the center $(2, 1, 2)$ and the radius 7.

The solution of the simultaneous equations of the line and the sphere may be obtained by finding y and z in terms of x from the equations of the line and substituting these values in the equation of the sphere.

20. Every section of a conicoid (§ 288) made by a plane parallel to one of the coordinate planes is a conic.

Since any plane could be taken as the xy plane, this theorem shows that every plane section of a conicoid is a conic.

SUPPLEMENT

I. The Quadratic Equation $ax^2 + bx + c = 0$

1. Roots. The roots are $x = \frac{-b \pm \sqrt{b^2 - 4ac}}{2a}$.

2. Sum of Roots. The sum of the roots is $-\frac{b}{a}$.

3. Product of Roots. The product of the roots is $\frac{c}{a}$.

4. Real, Imaginary, or Equal Roots. The roots are real and unequal if $b^2 - 4ac > 0$, equal if $b^2 - 4ac = 0$, and imaginary if $b^2 - 4ac < 0$.

5. Zero Roots. One root of the equation is 0 if and only if $c = 0$, the equation in this case being $ax^2 + bx = 0$; both roots are 0 if and only if $c = 0$ and $b = 0$.

6. Infinite Roots. One root of the equation increases without limit, or, as we say, is infinite, if and only if $a \rightarrow 0$; both roots are infinite if and only if $a \rightarrow 0$ and $b \rightarrow 0$.

This may be shown by considering the equation

$$ax^2 + bx + c = 0. \tag{1}$$

If we let $x = \frac{1}{y}$ and clear of fractions, equation (1) becomes

$$cy^2 + by + a = 0. \tag{2}$$

If one value of y in (2) is very small, then one value of x in (1) is very large, since $x = \frac{1}{y}$. Thus, when $y \rightarrow 0$, x is infinite. But the condition that one value of y approaches 0 as a limit is that $a \rightarrow 0$ (§ 5 above). Therefore x is infinite when $a \rightarrow 0$, and the same method shows the conditions for which both roots are infinite.

II. Logarithms of Numbers from 0 to 99

N	0	1	2	3	4	5	6	7	8	9
1	000	041	079	114	146	176	204	230	255	279
2	301	322	342	362	380	398	415	431	447	462
3	477	491	505	519	531	544	556	568	580	591
4	602	613	623	633	643	653	663	672	681	690
5	699	708	716	724	732	740	748	756	763	771
6	778	785	792	799	806	813	820	826	833	839
7	845	851	857	863	869	875	881	886	892	898
8	903	908	914	919	924	929	934	940	944	949
9	954	959	964	968	973	978	982	987	991	996

The mantissas of the logarithms of numbers from 1 to 9 are given in the first column. Decimal points are understood before all mantissas. To find log 84, write 1 as the characteristic and read the mantissa after 8 and under 4; thus, log 84 = 1.924.

III. Square Roots of Numbers from 0 to 99

N	0	1	2	3	4	5	6	7	8	9
0	0.00	1.00	1.41	1.73	2.00	2.24	2.45	2.65	2.83	3.00
1	3.16	3.32	3.46	3.61	3.74	3.87	4.00	4.12	4.24	4.36
2	4.47	4.58	4.69	4.80	4.90	5.00	5.10	5.20	5.29	5.39
3	5.48	5.57	5.66	5.74	5.83	5.92	6.00	6.08	6.16	6.24
4	6.32	6.40	6.48	6.56	6.63	6.71	6.78	6.86	6.93	7.00
5	7.07	7.14	7.21	7.28	7.35	7.42	7.48	7.55	7.62	7.68
6	7.75	7.81	7.87	7.94	8.00	8.06	8.12	8.19	8.25	8.31
7	8.37	8.43	8.49	8.54	8.60	8.66	8.72	8.77	8.83	8.89
8	8.94	9.00	9.06	9.11	9.17	9.22	9.27	9.33	9.38	9.43
9	9.49	9.54	9.59	9.64	9.70	9.75	9.80	9.85	9.90	9.95

The square roots of numbers from 0 to 9 are in the top row of the table. Thus the square root of 6 is 2.45.

To find the square root of a number having two digits, find the tens' digit in the column under N, the units' digit in the row to the right of N, and take from the table the number which corresponds. Thus $\sqrt{39} = 6.24$. The roots are given to the nearest hundredth.

IV. Values of Trigonometric Functions

Angle	sin	cos	tan	cot	
1°	0.017	1.000	0.017	57.29	89°
2°	.035	0.999	.035	28.64	88°
3°	.052	.999	.052	19.08	87°
4°	.070	.998	.070	14.30	86°
5°	.087	.996	.087	11.43	85°
10°	.174	.985	.176	5.67	80°
15°	.259	.966	.268	3.73	75°
20°	.342	.940	.364	2.75	70°
25°	.423	.906	.466	2.14	65°
30°	.500	.866	.577	1.73	60°
35°	.574	.819	.700	1.43	55°
40°	.643	.766	.839	1.19	50°
45°	.707	.707	1.000	1.00	45°
	cos	sin	cot	tan	Angle

V. Radian and Mil Measure for Angles

In a circle whose radius is r units of length, let AB be an arc whose length is one radius. Then the angle AOB at the center is called a *radian*, and 0.001 of a radian is called a *mil.*

An angle of 2 radians obviously cuts off an arc of length $2r$; an angle of k radians cuts off an arc whose length l is given by the equation

$$l = kr.$$

Since r is contained 2π times in the circumference, the entire angle about O is 2π radians; that is, a little more than 6 radians, and

$$2\pi \text{ radians} = 360°.$$

Hence, also, π radians $= 180°$; $\frac{\pi}{2}$ radians $= 90°$; $\frac{\pi}{3}$ radians $= 60°$; 1 radian $= 57.2958°$; and 1 mil $= 0.057°$, approximately.

VI. Trigonometric Formulas

$$\sin A = \frac{1}{\csc A} \qquad \cos A = \frac{1}{\sec A}$$

$$\tan A = \frac{\sin A}{\cos A} \qquad \tan A = \frac{1}{\cot A}$$

$$\sin^2 A + \cos^2 A = 1 \qquad 1 + \tan^2 A = \sec^2 A$$

$$\sin(180° - A) = \sin A \qquad \sin(-A) = -\sin A$$

$$\cos(180° - A) = -\cos A \qquad \cos(-A) = \cos A$$

$$\tan(180° - A) = -\tan A \qquad \tan(-A) = -\tan A$$

$$\sin 2A = 2 \sin A \cos A \qquad \cos 2A = \cos^2 A - \sin^2 A$$

$$\tan 2A = \frac{2 \tan A}{1 - \tan^2 A} \qquad \cot 2A = \frac{\cot^2 A - 1}{2 \cot A}$$

$$\sin(A \pm B) = \sin A \cos B \pm \cos A \sin B$$

$$\cos(A \pm B) = \cos A \cos B \mp \sin A \sin B$$

$$\tan(A \pm B) = \frac{\tan A \pm \tan B}{1 \mp \tan A \tan B}$$

$$\cot(A \pm B) = \frac{\cot A \cot B \mp 1}{\cot B \pm \cot A}$$

$$\sin \tfrac{1}{2}A = \pm\sqrt{\frac{1 - \cos A}{2}} \qquad \tan \tfrac{1}{2}A = \pm\sqrt{\frac{1 - \cos A}{1 + \cos A}}$$

$$\cos \tfrac{1}{2}A = \pm\sqrt{\frac{1 + \cos A}{2}} \qquad \cot \tfrac{1}{2}A = \pm\sqrt{\frac{1 + \cos A}{1 - \cos A}}$$

$$\sin A = 2 \sin \tfrac{1}{2}A \cos \tfrac{1}{2}A \qquad \cos A = \cos^2 \tfrac{1}{2}A - \sin^2 \tfrac{1}{2}A$$

$$\tan A = \frac{2 \tan \frac{1}{2}A}{1 - \tan^2 \frac{1}{2}A} \qquad \cot A = \frac{\cot^2 \frac{1}{2}A - 1}{2 \cot \frac{1}{2}A}$$

$$\frac{\sin A + \sin B}{\sin A - \sin B} = \frac{\tan \frac{1}{2}(A + B)}{\tan \frac{1}{2}(A - B)}$$

NOTE ON THE HISTORY OF ANALYTIC GEOMETRY

Since a considerable part of analytic geometry is concerned with conic sections, it is of interest to observe that these curves were known to the Greeks, at least from the fourth century B.C. We know that Menæchmus (*c.* 350 B.C.) may have written upon them, that for a long time they were called Menæchmian triads, that Aristæus, Euclid (*c.* 300 B.C.), and Archimedes (*c.* 250 B.C.) all contributed to the theory, and that Apollonius (*c.* 225 B.C.) wrote an elaborate treatise on conics. So complete was this work of Apollonius that it seemed to mathematicians for about eighteen hundred years to have exhausted the subject, and very few additional discoveries were made. Most of the propositions given today in elementary treatises on analytic geometry are to be found in the writings of Apollonius, and some idea of coordinates may also be seen in certain passages of his work. The study of conics by ancient and medieval writers was, however, of a purely geometric type, the proofs of the propositions being of the same general nature as the proofs of the propositions in Euclid's celebrated work on geometry.

DESCARTES

During the Middle Ages there appeared two works entitled *De Latitudinibus Formarum* and *Tractatus de Uniformitate et Difformitate Intensionum*, both written by Nicolas Oresme (*c.* 1323–1382), a teacher in the Collège de Navarre at Paris. In these works Oresme locates points by means of coördinates, somewhat as places are located on a map by means of latitude and longitude. In geographic work this method of location had been used long before by such Greek scholars as Hipparchus, Marinus of Tyre, Ptolemy, and Heron.

FERMAT

While there was some revival of an interest in conics in the works of such writers as Kepler (1571–1630), it was not until Descartes (Latin, Cartesius, 1596–1650) published his epoch-making little work *La Géométrie*, in 1637, that the fundamental ideas of analytic geometry were laid before the mathematical world. It is true that Fermat (*c.* 1601–1665), as shown by his correspondence with contemporary scholars, had already conceived the idea that the properties of a curve could be deduced from its equation, but, as in the case of his contributions to the theory of numbers, he published nothing upon the subject, and so the work of Descartes stands as the first to make known the ideas of analytic geometry. Although Descartes had the idea of an analytic geometry of three dimensions, his work is confined entirely to the plane, and it was left for men like Parent (1700), and especially Clairaut (1731) and Euler (1760), to extend the theory to curves of double curvature and to surfaces.

INDEX

www.ingramcontent.com/pod-product-compliance
Lightning Source LLC
LaVergne TN
LVHW091638100826
845152LV00005B/89

* 9 7 8 1 4 1 8 1 6 6 2 5 0 *